Hochschulschriften

Institut für Systembiotechnologie
Universität des Saarlandes

Herausgegeben von Prof. Dr. Christoph Wittmann

Band 5

Cuvillier-Verlag
Göttingen, Deutschland

Herausgeber
Univ.-Prof. Dr. Christoph Wittmann
Institut für Systembiotechnologie
Universität des Saarlandes
Campus A1.5, 66123 Saarbrücken
www.iSBio.de

Hinweis: Obgleich alle Anstrengungen unternommen wurden, um richtige und aktuelle Angaben in diesem Werk zum Ausdruck zu bringen, übernehmen weder der Herausgeber, noch der Autor oder andere an der Arbeit beteiligten Personen eine Verantwortung für fehlerhafte Angaben oder deren Folgen. Eventuelle Berichtigungen können erst in der nächsten Auflage berücksichtigt werden.

Bibliographische Informationen der Deutschen Nationalbibliothek
Die Deutsche Nationalbibliothek verzeichnet diese Publikation in der Deutschen Nationalbibliographie; detaillierte bibliographische Daten sind im Internet über *http://dnb.d-nb.de* abrufbar.
1. Aufl. – Göttingen: Cuvillier, 2019

Nonnenstieg 8, 37075 Göttingen
Telefon: 0551-54724-0
Telefax: 0551-54724-21
www.cuvillier.de

1. Auflage, 2019
Gedruckt auf umweltfreundlichem, säurefreiem Papier aus nachhaltiger Forstwirtschaft.

ISBN 978-3-7369-9971-8
eISBN 978-3-7369-8971-9
ISSN 2199-7756

Development and Application of a Method for Quantitative Metabolome Analysis of Various Production Strains

Dissertation

zur Erlangung des Grades

des Doktors der Ingenieurwissenschaften

der Naturwissenschaftlich-Technischen Fakultät

der Universität des Saarlandes

von

Georg Martin Richter

Saarbrücken

2018

Tag des Kolloquiums:	28.09.2018
Dekan:	Prof. Dr. G. Kickelbick
Vorsitz:	Prof. Dr. G.-W. Kohring
Berichterstatter:	Prof. Dr. C. Wittmann
	Prof. Dr. E. Heinzle
Akademischer Mitarbeiter:	Dr. J. Neunzig

Participants – Colleagues

Partial results were determined in collaboration with colleagues as referred to in this thesis.

Cultivations of *Y. pseudotuberculosis* were performed in collaboration with Dr.-Ing. René Bücker, Institute of Systems Biotechnology, Universität des Saarlandes.

Cultivations of *B. megaterium* were conducted in collaboration with Dr.-Ing. Thibault Godard, Institute of Biochemical Engineering, TU Braunschweig.

Cultivations of *D. shibae* were performed in collaboration with Dr.-Ing. Annekathrin Bartsch, Institute of Systems Biotechnology, Universität des Saarlandes.

Participants – Bachelor students

Partial results published in this work were determined in collaboration with M. Sc. Biotech. Christina Engel. During her bachelor thesis at the Institute of Biochemical Engineering, TU Braunschweig, Ms. Engel compared the extraction efficiency of boiling ethanol and cold acidic acetonitrile methanol mixture. Furthermore, she produced the $U^{13}C$ labeled internal standard, which was used for precise quantitative measurements.

Danksagung

An der Entstehung dieser Arbeit war eine Vielzahl an Menschen direkt und indirekt beteiligt, bei denen ich mich ganz herzlich bedanken möchte.

Ein besonderer Dank gilt meinem Doktorvater Prof. Dr. Christoph Wittmann für die Bereitstellung des spannenden Themas und die stets gute Betreuung der Arbeit, auch unter widrigsten Bedingungen. Die entspannte Atmosphäre bei den gemeinsamen Besprechungen und die konstruktiven Anregungen haben diese Arbeit in vielerlei Hinsicht bereichert. Vielen Dank auch für das Vertrauen das du mir entgegengebracht hast in dem du mich meine Arbeit in Braunschweig hast zu Ende bringen lassen. Ich weiß, dass die Betreuung meiner Arbeit dadurch nicht einfacher geworden ist!

Mein Dank gilt weiterhin den Professoren Elmar Heinzle und Gert-Wieland Kohring für die Bereitschaft zur Übernahme des Koreferats bzw. des Prüfungsvorsitzes. Vielen Dank auch an Herrn Dr. Björn Becker für den Prüfungsbeisitz als akademischer Mitarbeiter.

Des Weiteren möchte ich mich bei Herrn Prof. Dr. Rainer Krull bedanken, der bedingungslos den letzten verbliebenen Saarbrückener Exilanten im Institut in Braunschweig Unterschlupf gewährt hat und immer ein offenes Ohr für die alltäglichen Probleme eines Doktoranden hatte.

Danken möchte ich auch Annekatrin Bartsch, Rene Bücker und Thibault Godard, deren Wissen über die Eigenheiten von *D. shibae*, *Y. pseudotuberculosis* und *B. megaterium* erst ermöglicht haben, dass diese Arbeit einen ausführlichen Überblick über das mikrobielle Metabolom verschiedenster Stämme geben kann.

Bedanken möchte ich mich auch bei den Mitarbeitern des Institutes für Bioverfahrenstechnik der TU Braunschweig und den Mitarbeitern des Institutes für Systembiotechnologie der Universität des Saarlandes. Die angenehme Arbeitsatmosphäre hat es immer leicht gemacht am morgen früh im Labor zu erscheinen um es abends erschöpft wieder zu verlassen. Es gab kein Problem, dass nicht mit Hilfe der Schwarmintelligenz angegangen werden konnte, keine Nachtschicht, die nicht bei Computerspielen kürzer wurde (Danke, André) und keinen Frust, der nicht beim gemeinsamen Feierabendbier oder Sport kleiner wurde. Vielen Dank euch allen, ihr habt die gemeinsame Zeit unvergesslich werden lassen. Ein spezieller Dank hier noch an meinen Bürokollegen Arne, der alle meine Macken im Büro mit stoischer Ruhe ertragen hat um sie dann auch noch im gemeinsamen USA-Urlaub zu genießen.

Außerdem möchte ich mich bei „meiner" Studentin Christina Engel bedanken, die sich selbst durch den eigenen Geburtstag nicht von einer Nachtschicht mit mir und *C. glutamicum* abhalten ließ. Deine Arbeit und die vielen lustigen gemeinsamen Stunden im Labor haben diese Arbeit maßgeblich bereichert.

Allen Freunden, die mich immer wieder aus dem Doktorandenalltag geholt haben möchte ich für die vielen netten gemeinsamen Stunden danken. Die gemeinsamen Urlaube in Dänemark, die vielen gemeinsamen Spieleabende, Videoabende am Dienstag bei uns auf dem Sofa und die Zockrunde am Donnerstag haben dafür gesorgt, dass ich immer wieder ausgeglichen ins Labor zurückkehren konnte.

Ganz besonders möchte ich Sarah danken, die mich immer voll und ganz unterstützt hat und ohne die diese Arbeit mit Sicherheit ganz anders ausgesehen hätte. Vielen Dank für dafür, dass du jeden Tag für mich da bist und es einfach immer schaffst mich zum Lachen zu bringen.

Ein letzter Dank geht an meine Familie, die mich bis hierher gebracht hat und mich in Studium und Promotion immer unterstützt hat. An meinen Vater der mich immer wieder angetrieben hat. An meine Mutter für ihr offenes Ohr, ihre Anregungen und die vielen Stunden zwischendurch, wenn es mal nicht um die Dissertation ging. Danke auch an Björn, ohne deine Hilfe bei Simulationen und Grafiken säße ich wahrscheinlich immer noch vor dem Rechner.

Vielen Dank euch allen!

Table of Contents

I. Summary

Recent technical improvements in the area of mass spectrometry enabled quantitative measurements of small molecules. The present work used these advancements to develop a method for the quantification of metabolites of the central carbon metabolism in the nanomolar range. Sample analysis by tandem mass spectrometry hyphenated with liquid chromatography enabled measurements of 33 metabolites in 25 min. Performed measurements were validated by thermodynamic and energetic constraints, which proved that datasets were of high quality. Combination of metabolomics and fluxomics approaches enabled a holistic and systems-oriented view during analysis of the metabolic datasets. The new method was first applied to identify changes in the energy charge of *E. coli* during different cultivation modes. Analysis showed that the adenylate energy charge was actively controlled by secretion or synthesis of adenylate phosphates. Thus, a strong imbalance between energy generation and consumption was necessary to distort the energy charge permanently. The novel technique was further used to identify changes in the central carbon metabolism of microorganisms as a consequence of genetic modifications, stress inducing cultivation conditions and changes in carbon source. Surprisingly, genetic modifications and stress inducing cultivation conditions resulted only in minor changes of intracellular metabolite levels. Hence, it seems that microorganisms put great efforts into the homeostasis of metabolite ratios and fluxes.

II. Zusammenfassung

Neue Entwicklungen im Bereich der Massenspektroskopie ermöglichen quantitative Messungen von Molekülen mit sehr kleinen Massen. In der vorliegenden Arbeit wurden diese Neuentwicklungen genutzt um Metabolite des zentralen Kohlenstoffwechsels im nanomolaren Bereich nachzuweisen. Die Kopplung von Tandem-Massenspektroskopie und Flüssigchromatographie ermöglichte den Nachweis von 33 Metaboliten in 25 min. So gewonnene Datensätze wurden anschließen mittels Thermodynamik validiert, wodurch die hohe Qualität der Datensätze deutlich wurde. Bei der Analyse der Datensätze wurden dann Ergebnisse aus Metabolom- und Fluxomforschung kombiniert um einen ganzheitlichen Ansatz zu schaffen. Die neue Technik wurde zuerst angewandt um den Einfluss verschiedener Kultivierungsverfahren auf das Energielevel von *E. coli* zu untersuchen. Die Untersuchungen zeigten, dass *E. coli* seinen Energielevel aktiv durch die Sekretion und Synthese von Energiemetaboliten steuern kann. Nur extreme Ungleichgewichte bei Verbrauch und Generierung von Energie konnten den Energiehaushalt nachhaltig stören. Weitere Messungen untersuchten den Einfluss von genetischen Veränderungen, Stress und unterschiedlichen C-Quellen auf den zentralen Kohlenstoffwechsel von Mikroorganismen. Überraschenderweise führten weder genetische Veränderungen noch Stress zu starken Veränderungen der Metabolitkonzentrationen. Dies zeigt das Mikroorganismen viel Aufwand in die Homöostase von Metabolit-Verhältnissen und intrazellulären Flüsse stecken.

1 Introduction

Metabolic pathways are the very essence of life and the understanding of their functioning and their regulation has been a major goal of the natural sciences, since the discovery of metabolism itself. Interestingly, central carbon metabolism is highly conserved in most living organisms due to its efficiency (Smith & Morowitz 2004; Ebenhöh & Heinrich 2001). It channels various nutrients through pathways to provide energy and reducing power for cell maintenance and reproduction as well as precursor metabolites for biomass formation (Noor et al. 2010). As such, this core piece of metabolism is in the heart of biotechnology. Elucidation of central carbon metabolism during the past decades (Entner & Doudoroff 1952; Gunsalus et al. 1955; Krebs & Johnson 1980; Barnett 2003) laid the foundation for modern biotechnology. The meanwhile even deeper understanding of metabolism and its regulatory elements is a valuable knowledge base to enhance bio-production processes, as it enables the rational design of metabolic pathways towards high titers and yields of desired products of choice (Akinterinwa et al. 2008; Buchinger et al. 2009; Becker et al. 2011; Becker & Wittmann 2012b). Particularly, sophisticated methods, which allow the detection of imbalances in levels of energy or precursor metabolites and the identification of even unknown cellular intermediates, appear most useful to identify genetic targets for tailor-made modifications of microorganisms towards improved performance (Lee et al. 2005; Wendisch et al. 2006; Becker et al. 2007).

More fundamentally, the exact quantification of intracellular metabolites allows describing and modeling of the complex underlying metabolic pathway networks. Such systems-oriented approaches heavily rely on suitable technologies, which provide accurate data of the studied system. Experimental techniques that provide such systems wide insights are therefore well suited to collect quantitative data to construct, adjust and validate systems biology models. These techniques are referred to as "omics" and involve genomics, transcriptomics, proteomics, fluxomics and metabolomics according to their target analytes (Romualdi & Gerolamo 2009). Hence, the field of metabolomics intends to study the profiles of metabolites, which are products of specific cellular processes (Jordan et al. 2009) and thereby provides a snapshot of cellular physiology at a certain point of time. As a subcategory, metabonomics is by definition "the quantitative measurement of the dynamic multiparametric metabolic response of living systems to pathophysiological stimuli or genetic modifications" (Nicholson 2006).

The recent decades have seen a tremendous progress in the field of metabolite analysis. In the pioneering days, colorimetric, manometric and enzymatic assays were employed for identification and quantification of metabolites (Entner & Doudoroff 1952; Krebs & Johnson 1980). The elucidation of the central pathways of metabolism by utilization of these methods took an enormous effort, as such techniques required large amounts of the analytes of interest and multiple steps of purification prior to the actual measurement. The coupling of metabolite analysis to chromatographic separation processes, such as thin-layer chromatography (TLC) and high performance liquid chromatography (HPLC), led to drastically lowered detection limits, thus displaying a milestone in the analysis of sample mixtures (Touchstone 1993). This was even improved by coupling of mass spectrometry (MS) to chromatographic separation processes. The use of highly sensitive tandem mass spectrometry for detection led to a further decrease of detection limits, while simultaneously allowing quantification of complex metabolite mixtures, due to the identification of specific metabolite fragments by their mass to charge ratio (Luo et al. 2007; Balcke et al. 2011). In addition, mass spectrometry allows distinguishing between different isotopically labeled versions of a metabolite, as a consequence of the change in mass, thereby enabling approaches like ^{13}C metabolic flux analysis (Wittmann & Heinzle 1999; Dauner & Sauer 2000). Most recently, the utilization of ion mobility mass spectrometry enables the determination the ratios of isomers, which are inseparable by chromatography (Far et al. 2014). These recent instrumental developments now provide an excellent basis to perform metabolite analysis.

Basically, it has become relatively easy to generate larger datasets of intracellular metabolite concentrations. However, remaining questions revolve around the challenge that the measured levels indeed reflect the studied *in vivo* system (Bolten et al. 2007). Particularly, the quality of metabolic datasets is strongly dependent on the preceding steps of sampling, sample treatment and processing. It is crucial to stop the metabolism instantaneously due to high turnover rates and small pool sizes of the metabolites of interest, which might otherwise be distorted (Wittmann et al. 2005). Furthermore, the avoidance of potential unwanted side effects of the sample treatment, such as metabolite leakage (Bolten et al. 2007), co-precipitation (Zakhartsev et al. 2015) and the degradation or interconversion of metabolites during the extraction, has to be considered. Combined with the largely varying physico-chemical properties of metabolites and differences in concentrations of several orders of magnitudes these problems still pose as a serious challenge when developing a reliable quantitative method for metabolite analysis.

2 Objectives

The aim of the present study was to develop a method for the quantification of intermediates from central carbon metabolism via liquid chromatography coupled with tandem mass spectrometry, suitable to be applied to a variety of different microorganisms. Among the analytes of interest were metabolites of the glycolysis, the Entner-Doudoroff (ED) pathway, the pentose phosphate (PP) pathway and the tricarboxylic acid (TCA) cycle, as well as energy metabolites and redox equivalents. Preferably, the analysis of cellular extracts should be faster as commonly reported time periods of up to 90 minutes and exhibit a high-resolution capacity for the complex cellular extracts of interest, which should be aimed for by optimization of the liquid chromatography and the coupling to mass spectrometric detection. The developed method should generate reproducible results and combine high separation efficiency with high sensitivity and short analysis times. Starting initially with standard mixtures from synthetic compounds the approach should be transferred to living cells. This should involve the estimation of the limits of detection (LOD) and quantification (LOQ) in the presence of relevant biological matrices. Additionally, different protocols for sampling and metabolite extraction should be applied and validated. Careful inspection of the obtained datasets from thermodynamic and energetic perspective should then provide a clear basis for validation.

Finally, the novel quantitative method should be applied to address relevant biological questions. Hence, the central carbon metabolome of seven biotechnological relevant Gram-negative and Gram-positive microbial strains should be studied under different growth conditions. The gained insight should then be combined with results from metabolic flux analysis to investigate the impact of genetic modifications, environmental stress or changes of substrate on selected strains.

3 Theoretical Background

3.1 Quantitative Measurements in Microbial Metabolomics

Generally, metabolomics investigates the endo- and exo-metabolome by qualitative, semi-quantitative and quantitative approaches respectively. Hereby, quantitative analysis gives the most accurate picture. In contrast to semi-quantitative approaches (Zamboni et al. 2008), quantitative datasets can be validated by thermodynamic inspection and used for kinetic and metabolic modeling (de Jonge et al. 2014; Wiechert & Noack 2011). Metabolomics itself is a relatively young component in the field of systems biology. Since the year 2000 more and more studies involve metabolomic measurements, indicating a fast-growing interest in this field (Figure 3.1). Obviously, metabolomics is becoming a valuable tool of research in the fields of metabolic engineering (Trethewey 2004; Toya & Shimizu 2013) and bioprocess optimization (Sonntag et al. 2011) over the last decade. However, compared to e.g. transcriptomics, genomics and proteomics, the number of metabolomic studies seems still rather low (Kohlstedt et al. 2010). Particularly, open questions remain and involve suitable protocols for sampling, quenching and extraction procedures as well as analytical methods to guarantee validity and comparability of metabolic datasets. Without doubt, metabolome analysis remains an analytical challenge (van der Werf et al. 2007).

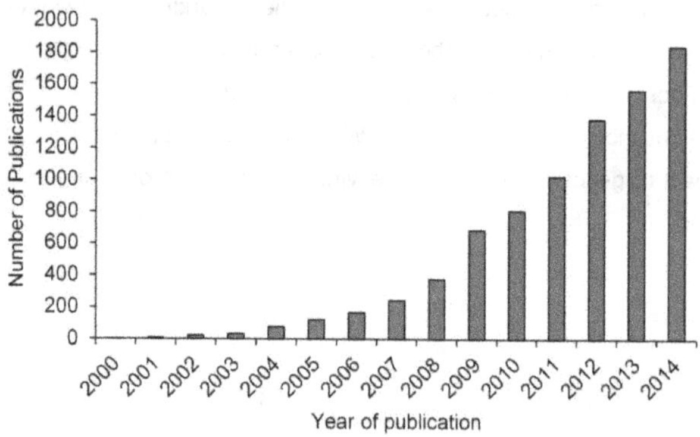

Figure 3.1: Increase in the number of publications in the field of metabolomics over the past 14 years according to the PubMed database (2015).

A powerful and robust quantitative method for the measurement of intracellular metabolites has to meet several key criteria. The method should provide high resolution, short measurement time and high reproducibility. The first challenge, which has to be overcome to achieve this goal, is the complexity of biological samples, which comprise multiple compounds with great variance in concentration and chemical properties. As example, the metabolome of

E. coli consists out of about 1000 different metabolites, whereby intracellular concentrations range from picomolar to millimolar (Dunn & Ellis 2005; Feist et al. 2007). Thus, the applied analytical instruments have to enable high separation efficiency, while at the same time providing a high dynamic detection range.

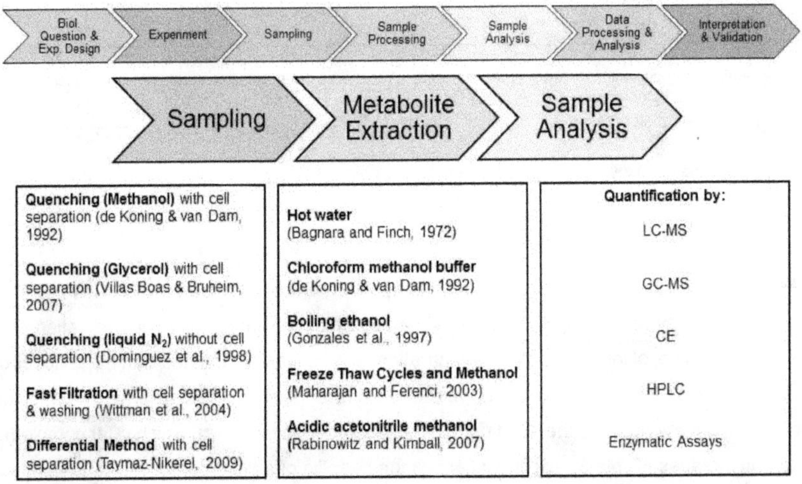

Figure 3.2: Standard workflow of a metabolomics experiment with emphasis on critical choices during method development.

Immediately emerging from the general metabolomics workflow (Figure 3.2) the establishment of reproducible sampling and extraction methods becomes mandatory. Turnover rates of metabolites of the central carbon metabolism are in the range of seconds to sub-seconds, making it crucial to stop all enzyme activities immediately during sampling, while at the same time avoiding metabolite leakage (Wittmann et al. 2004; van Gulik 2010). Subsequently, an extraction method has to be chosen, which combines reproducibility and extraction efficacy, while simultaneously avoiding degradation of metabolites as a consequence of harsh extraction conditions (Villas-Bôas et al. 2005; Canelas et al. 2009). Furthermore, the application of a uniformly ^{13}C labeled internal standard, prior to the extraction process, has been proven a powerful tool to compensate for effects of biological matrices in the sample, as well as effects of degradation during the extraction process (Mashego et al. 2007).

Table 3.1: Strengths and weaknesses of the hyphenation of gas and liquid chromatography to mass spectrometry for metabolomics

	GC	LC
Advantages	- electron impact (EI) ionization highly reproducible and robust - high separation efficiency - spectral libraries available	- high variety of separation mechanisms enable detection of a wide range of analytes - applicable to non-volatile, thermally labile and polar compounds - electrospray-ionization (ESI) is a soft ionization type (no fragmentation)
Disadvantages	- limited applicability to non-volatile, thermally labile and polar compounds - derivatization necessary - fragmentation during the ionization step	- sensible to matrix dependent suppression effects - limited flexibility in eluent composition due to ESI

As stated above, microbial extracts contain up to 1000 different low molecular weight compounds, thereby requesting for efficient separation mechanisms prior to quantification of selected molecules of interest. Different separation techniques, hyphenated with mass spectrometry (MS), have emerged in quantitative metabolomics. Gas chromatography (GC) and liquid chromatography (LC) are the most prominent separation mechanism, which offer complementary advantages (Table 3.1). The GC unites the benefits of a wide compound range and high separation efficiency with the high reproducibility of electron impact (EI) ionization. As a consequence, a variety of spectral libraries can be found, which enable the easy identification of a huge variety of metabolites. Due to its high sensitivity, mass accuracy and reproducibility GC-MS has become a standard tool, which is widely applied to study fluxes in prokaryotic (Park et al. 1997) and eukaryotic cells (Christensen et al. 2000) or to elucidate biosynthetic pathways (Hellerstein et al. 1991; Kelleher et al. 1994). GC of nonvolatile or polar metabolites, however, requires an additional derivatization step to render these compounds amendable to gas phase separation, thus adding a potential bias due to incomplete derivatization (Büscher et al. 2009). Moreover, the subsequent vaporization hinders the successful measurement of thermally labile metabolites and the hard ionization techniques limit the gain of additional structural information.

However, these drawbacks can be overcome by coupling liquid chromatography to mass spectrometry. In contrast to GC-MS, LC coupled MS uses soft ionization techniques, such as electrospray ionization (ESI) and atmospheric-pressure chemical ionization (APCI). Thus, it is especially useful for the analysis of thermally labile, polar and nonvolatile compounds, of which many are found in the central carbon metabolism. Consequently, recent technical improvements in the coupling of liquid chromatography and mass spectrometry led to an increasing interest in this technology, due to its broad application potential in dynamic flux analysis (Oldiges et al. 2004; Kiefer et al. 2007) and high throughput screenings (Sauer 2004). A major drawback of LC methods is the susceptibility of separation and ionization quality to matrix effects. Fortunately, matrix effects can be lowered by addition of ion-pairing reagents, which prolong the retention times of metabolites on the column, thereby enhancing the separation process. As a consequence, the analytical method should be chosen carefully. The quantitative approach of targeted metabolomics focuses on the measurement of defined sets of intermediates or end products of metabolic pathways as in studies of drug induced metabolism (Li et al. 2012; Spaggiari et al. 2014). Further application fields are the identification of biomarkers in cells, tissues or body fluids (Alberice et al. 2013; Quinones & Kaddurah-Daouk 2009) and the utilization in forensic sciences. Here, the LC-MS is used to identify synthetic drugs in blood, urine or hair samples which cannot be identified by GC-MS (Thevis et al. 2008; Maurer 2005; Broecker et al. 2012) or to detect trace levels of chemical warfare agents (Hayes et al. 2004) and explosives (DeTata et al. 2013) due to its high sensitivity.

3.2 Tandem Mass Spectrometry hyphenated with Liquid Chromatography

3.2.1 Liquid Chromatography for Metabolite Separation

As stated above, the coupling of liquid chromatography to mass spectrometers allows the analysis of nonvolatile, polar, charged and thermally labile metabolites. Meanwhile, the development of ultra-high performance liquid chromatography (UHPLC) enables shorter analysis times (Plumb et al. 2005), while also improving chromatographic resolution (Guillarme et al. 2010) and peak capacity (Wilson et al. 2005). Liquid chromatography offers different kinds of stationary phases providing a highly versatile platform for the separation of chemicals with different properties. To date, reversed phase (RP) columns, combined with a gradient elution, represent the most favored separation mode (Theodoridis et al. 2012). However, when applied to polar or ionic metabolites, such as organic acids and amino acids, it is important to use additional separation mechanisms. One possibility is the use of specific stationary phases like hydrophilic interaction liquid chromatography (HILIC) (Spagou et al. 2011), aqueous normal phase chromatography (ANP) (Pesek & Matyska 2005; Callahan et al. 2009) or reversed phase pentafluorophenylpropyl (PFPP) columns (Yang et al. 2010), respectively. HILIC columns combine the separation mechanisms of ion chromatography, normal phase and reversed phase and perform well in the analysis of uncharged highly hydrophilic and amphiphilic metabolites. The disadvantages of this technique are time consuming re-equilibration, column bleeding, decreasing resolution and instable retention times of metabolites (Walker et al. 2012; Snyder et al. 2010).

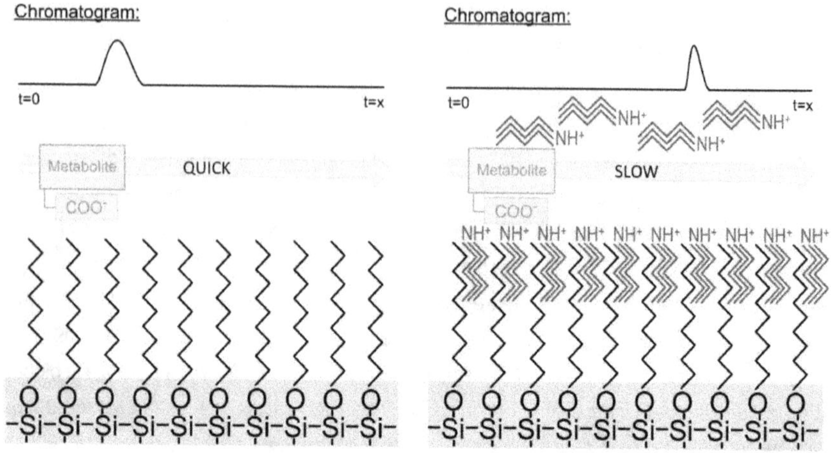

Figure 3.3: Working principle of reversed phase ion pairing (RP-IP) chromatography. Ionic compounds are badly retained by reversed phase chromatography. Ion pairing agents consist of an ionic end and a hydrophobic tail. Thus, the ion pairing reagent acts as mediator between hydrophobic stationary phase and the ionic analytes, thereby enhancing the interaction between the compound of interest and stationary phase. As a result, ionic compounds are retained much more efficient, resulting in a higher separation efficiency of the chromatographic method.

An interesting alternative is the use of ion pairing reagents (IP-LC) (Balcke et al. 2011), such as tributylamine (TBA) (Figure 3.3). Ion pairing reagents are often used to prolong the retention times of anionic compounds on HPLC-columns, in order to achieve a better separation and peak resolution. The acidic groups of the metabolites form ion pairs with the amine groups of the alkyl amines at a neutral pH, while the butyl residues interact with the hydrophobic tails of the stationary phase. The longer the alkyl chain, the stronger the hydrophobicity gets, resulting in better resolution. It was shown, that the use of alkyl chains with optimal length is crucial to obtain good resolution and peak shape. Short alkyl chains, as found in triethylamine, do not provide sufficient resolution for separation of structural isomers like G6P/F6P, while longer alkyl chains, such as octylamine, (Huck et al. 2003), can lead to peak broadening and therefore lower sensitivity. However, non-volatile ion pair reagents can lead to ion suppression, source pollution and the generation of high background signals (Gustavsson et al. 2001; Holčapek et al. 2004).

3.2.2 Mass Spectrometry for Metabolite Detection

Mass spectrometers (MS) are sensitive and precise detectors, which lifted the analysis of complex biological samples to a new level (Yates 2001). Prominent examples involve the quantification of intracellular fluxes towards resolving complex cellular responses to environmental and genetic perturbation (Wittmann et al. 2007). In addition, specific pulse response experiments combined with labeling information enable the investigation of new metabolic pathways (Peyraud et al. 2009). Gained knowledge is subsequently applied in targeted strain improvement (Becker et al. 2005) and metabolic engineering (Becker et al. 2007). Summarizing, application of mass spectrometry in quantitative metabolomics has become one of the key technologies of the last decade (Wittmann 2007).

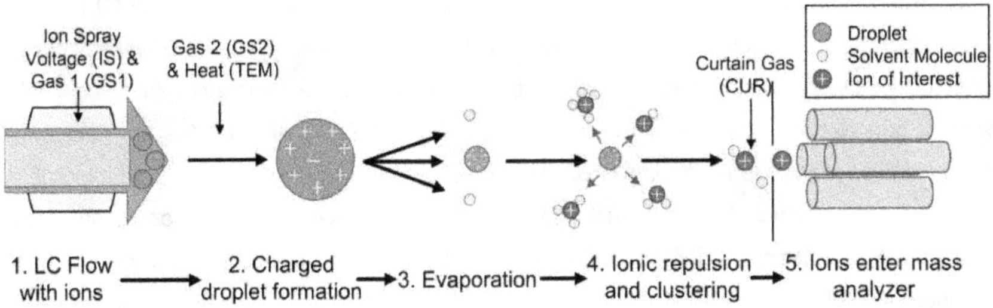

Figure 3.4: Working principle of Electrospray Ionization (ESI) operated in positive mode, used during metabolite quantification via LC-MS/MS measurements.

Due to its mild ionization conditions, the electrospray ionization (Figure 3.4) is preferred for the analysis of small molecules, which could otherwise fragment, before reaching the collision cell. It can be used in positive or negative mode, resulting in molecular ions which are protonated or deprotonated, respectively. As a consequence of the mild ionization conditions, the formation of adduct ions with constituents of mobile phase or sample matrices as well as dimerization of analytes have been reported (Schug & McNair 2002; Zhou et al. 2013). Formation of multiple charged ions is also possible, which leads to a drastic change in the mass to charge ratio. Both phenomena should therefore be taken into account, while determining the settings for the analysis of metabolites (Lynn et al. 2015). A minor problem that can be overcome by careful sample preparation is the sensitivity to contaminations by detergents or salts. With electron impact, atmospheric pressure chemical ionization, atmospheric pressure photo ionization or electro spray ionization, there are various different ionization types available for coupling of mass spectrometry to liquid chromatography, from which the latter is most suitable for metabolic profiling studies (Ho et al. 2003; Mortier et al. 2004).

The working principle of mass spectrometry allows differentiation of compounds, which are otherwise inseparable by conventional separation mechanisms, such as chromatography. The differentiation between coeluting substances according to their mass to charge ratio (*m/z*-ratio) has several advantages over more common identification by UV-spectra or refractive index. It enables a reduction of time needed per measurement, as complex chromatographic methods become unnecessary as long as the analytes of interest show differing masses. Several different technologies have evolved in the field of mass spectrometry, which should be chosen carefully according to the goal of the planned studies. Time of flight (TOF) and Fourier transform (FT) mass spectrometers, for example, show high resolution and mass accuracy, but require longer duty cycles to achieve the high resolutions, which potentially leads to a loss of information if these instruments are combined with UHPLC or other fast LC systems (Breitling et al. 2006; Makarov & Scigelova 2010). Accordingly, they have limited potential for the fast analysis of complex mixtures and a compromise has to be met between a good peak coverage and high resolution. In contrast, the utilization of tandem mass spectrometry generates highly specific fragmentation patterns, thus enhancing the signal to noise ratio and thereby sensitivity in complex samples (de Hoffmann 1996; Kuze et al. 2013). Additionally, the measurement of fragmentation patterns allows the quantitative measurement of mass or even positional isotopomers, which enables generation of highly resolved flux maps due to the gain of structural information (Polce et al. 1996).

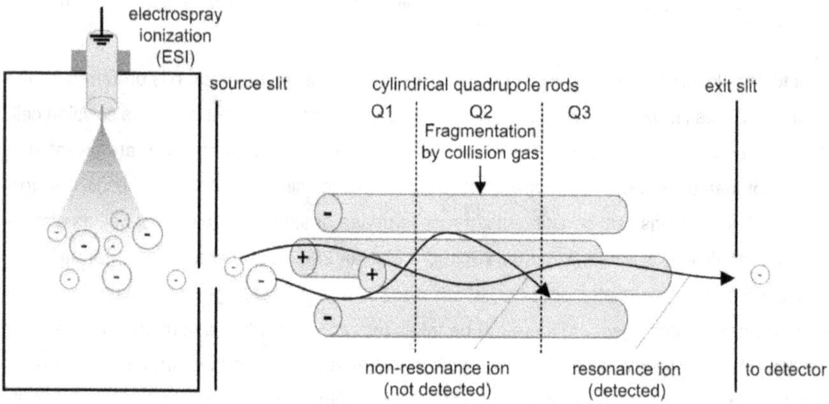

Figure 3.5: Schematic view of an LC-MS/MS system, combining Electrospray Ionization (ESI) and a triple stage quadrupole mass filter. Every Q-section is a single quadrupole

A quadrupole mass spectrometer consists of four metal rods aligned in parallel, with opposing rods being connected electrically. Ions are forced to travel down the quadrupole by an applied static electrical field. In addition, a radio frequency is applied to each of the opposing rod pairs. Depending on the radio frequency, only ions with a certain mass to charge ratio can pass the quadrupole on stable oscillating trajectories without colliding with the rods or even leaving the quadrupole. This allows the explicit selection of ions of interest or the scanning of a range of mass to charge ratios. In a triple quadrupole, the precursor ion is selected in the first quadrupole, fragmented in the second, while a specific fragment is selected in the third quadrupole, which is then conducted to the detector (Figure 3.5). To achieve high sensitivity, the fine adjustment of three parameters is of utmost importance. The first parameter is called the declustering potential (DP). The DP prevents ions from clustering together before entering the quadrupole. Optimization of the declustering potential guarantees a higher number of single molecules entering the mass spectrometer, thereby enhancing the overall sensitivity. The second parameter which needs optimization in a triple quadrupole is the collision energy (CE). Here the amount of energy is determined, which leads to a maximal yield of the fragment ion of choice. After the collision, the fragment ions have to be focused and accelerated into the third quadrupole. Here, the cell exit potential (CXP) needs to be optimized to allow a maximal number of fragment ions to enter the third quadrupole and subsequently the detector. The arrangement of three consecutive quadrupoles enables analysis of complex biological samples in various measurement modes (Figure 3.6).

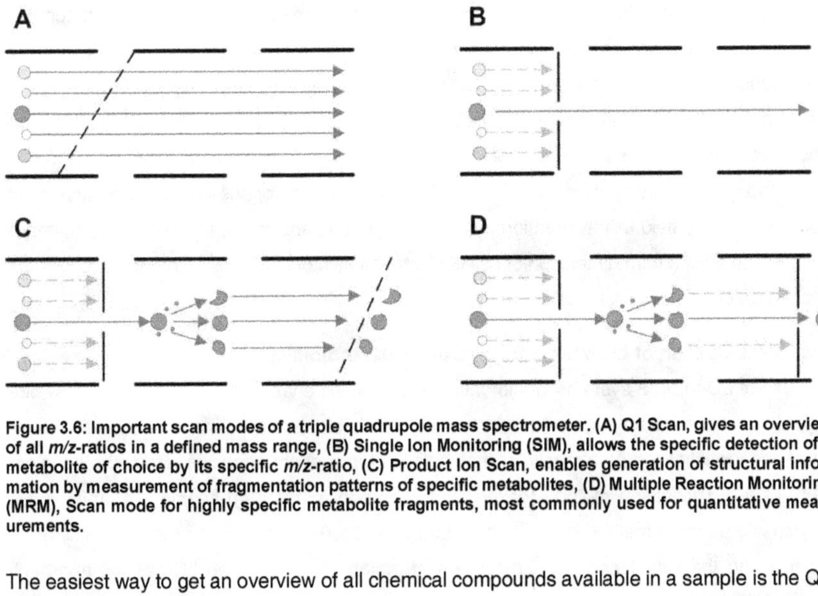

Figure 3.6: Important scan modes of a triple quadrupole mass spectrometer. (A) Q1 Scan, gives an overview of all *m/z*-ratios in a defined mass range, (B) Single Ion Monitoring (SIM), allows the specific detection of a metabolite of choice by its specific *m/z*-ratio, (C) Product Ion Scan, enables generation of structural information by measurement of fragmentation patterns of specific metabolites, (D) Multiple Reaction Monitoring (MRM), Scan mode for highly specific metabolite fragments, most commonly used for quantitative measurements.

The easiest way to get an overview of all chemical compounds available in a sample is the Q1 Scan. Here, a defined mass range is scanned and the abundance of all compounds which lie within the mass range is measured, thus enabling a quick overview which actual *m/z*-ratios can be found in the sample. If the *m/z*-ratio of the compound of choice is known, the single ion monitoring mode allows the specific measurement of the ion. The settings of the quadrupole allow only the specific ion to pass through the mass spectrometer and into the detector, while all others oscillate on unstable trajectories and subsequently leave the spectrometer. In the product ion scan a single ion is chosen in the first quadrupole, fragmented in the second and all resulting fragments are measured afterwards. This mode allows screening for fragment ions which are highly specific for the compounds of interest and additionally generates structural information of the measured substances. Multiple reaction monitoring (MRM) is heavily utilized in quantitative metabolomics. Here, a specific ion is selected in Q1, fragmented in Q2 and only specific fragment ions are stabilized in Q3 and allowed into the detector. The coupling of a specific molecular ion to its specific fragment ion is called transition. Measurement of these transitions is highly compound specific and therefore predestined for quantitative measurement of complex biological samples.

The selection via fragmentation patterns results in a robust, fast and sensitive analytical performance. Thus, mass spectrometry allows differentiation of compounds, which are otherwise inseparable, however even this technique has limitations if isomeric substances come into play. Isomeric compounds, which are not separable by chromatography, can be differentiated if the fragmentation patterns differ. However, substances like hexose- or pentose-phosphates tend to fragment in similar patterns. Here, a new approach to enable differentiation of isomeric compounds is the utilization of ion-mobility mass spectrometry as an additional separation mechanism after chromatography and before mass spectrometry (Far et al. 2014). This analytical technique enables the separation and identification of ionized compounds by their movement through a carrier buffer gas of which the flow direction opposes the ion motion in an applied electrical field. After retention time and mass to charge ratio, the separation by mass to size ratio adds a third dimension of selectivity to the field of LC-MS/MS, thereby allowing the separation of isomeric metabolites.

3.2.3 Application of LC-MS/MS for Microbial Metabolomics

The use of LC-MS/MS systems in microbial metabolomics enables the elucidation of new biological questions. As the high sensitivity and selectivity of these systems allows the measurement and identification of intracellular metabolites at lowest abundancies. Thus, the development of a quantification method for the central carbon metabolism seems a promising approach. To date the method developed by Luo et al. (2007) is one of the most promising approaches in the simultaneous quantification of metabolites. Here, an LC-MS/MS approach combined with ion pairing reagents was used to quantify 29 metabolites of the central carbon metabolism in a single run. Evaluation of different ion pairing reagents showed that application of tributylamine resulted in the best compromise between elution time, resolution and sensitivity. However, the run time of 90 minutes left the method hardly feasible for a high throughput approach. In addition, it was shown that application of a fully labeled cell extract, as an internal standard, results in a greatly enhanced reproducibility and resolution of metabolic measurements (Wu et al. 2005). Simultaneous extraction of labeled standard and unlabeled sample made elaborate recovery checks redundant, as the isotopomers react similar to the conditions during sample preparation. Hence, a constant ratio of unlabeled compound to labeled standard is maintained for every metabolite of interest independent of the preparation conditions. Furthermore, isotopomers are equally influenced by effects of the biological matrix, thus enabling easy identification of false positive peaks and shifts in retention time. Even though it is still a long way to the quantitative analysis of the whole metabolome, the coupling of liquid chromatography to a mass spectrometer is already used for a broad range of applications in microbial metabolomics. This is a consequence of its high versatility due to many possible combinations of different types of separation, ionization and detection mechanisms. LC-MS/MS measurements are already used to estimate the total concentration of metabolites in microorganisms (Bennett et al. 2009) or to supervise *in vivo* dynamics of metabolites as a reaction to sudden changes in substrate concentrations with pulse-response experiments (Buchholz et al. 2002; Wu et al. 2006; Yuan et al. 2009). These *in vivo* measurements, unlike *in vitro* estimations, include the effect of regulatory active substances inside the cell and can therefore be used to develop more accurate kinetic and metabolic models of cells (de Jonge et al. 2014; Wiechert & Noack 2011). Combined with the knowledge about metabolite pool sizes and regulatory effects, production strains with higher efficiencies or even new pathways can be constructed (Hasunuma et al. 2011). Taken together, these publications show the huge potential of quantitative metabolomics in systems biotechnology.

3.3 Sample Processing for Metabolite Quantification

Obviously, the analysis of intracellular metabolites requires a number of carefully optimized steps to ensure the acquisition of valid, precise and reproducible datasets (Figure 3.7). The high conversion rates of enzymes require a sampling method, which stops conversions as fast as possible. Additionally, the processing method should ensure complete and unbiased extraction of metabolites, while simultaneously avoiding further changes in metabolite concentration due to physical, chemical or biological effects.

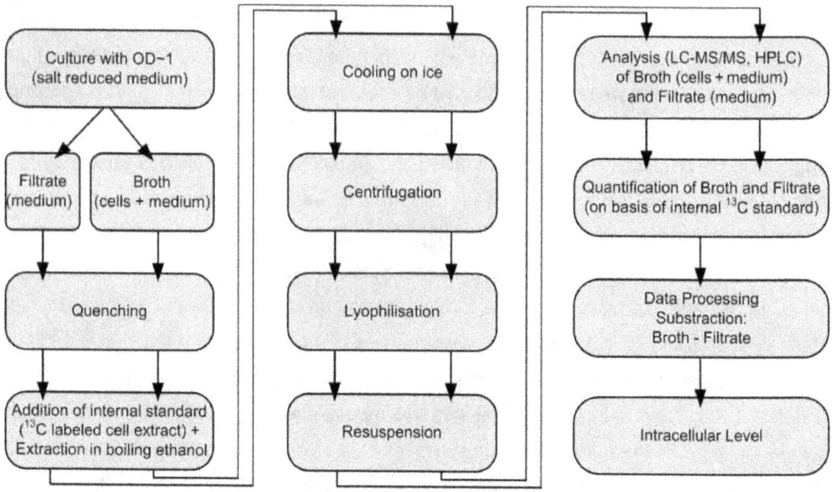

Figure 3.7: Paradigm of a workflow for sample processing during metabolite quantification.

3.3.1 Quenching of the Metabolome during Sampling

In metabolomics, quenching is used to stop all metabolite conversions upon sampling. If possible, even before separating the cells from the surrounding medium. One of the most popular procedures to quench samples is the use of cold aqueous 60% methanol (de Koning & van Dam 1992), while keeping the temperature below -10 °C (Gonzalez et al. 1997). This approach is particularly useful for eukaryotic microorganisms (Bolten & Wittmann 2008) but fails for most bacteria due to the fact that the used solvent damages the cell membrane and results in unspecific metabolite leakage out of the cell (Bolten et al. 2007; Winder et al. 2008).Yeasts and fungi for instance have a more stable cell wall and are therefore less affected by leakage than prokaryotes (Mashego et al. 2007; de Koning & van Dam 1992; Gonzalez et al. 1997).

Different variations of the methanol quenching process, like the addition of buffer salts (Winder et al. 2008), tricine (Castrillo et al. 2003) or glycine salts (Marcellin et al. 2009), proved to be ineffective to prevent cell leakage. An attractive approach to overcome these hurdles involves the separation of cells and medium by filtration and centrifugation, respectively (Bolten & Wittmann 2008). Hereby, it is important to use solutions of similar ionic strength during the washing step (Bolten et al. 2007). However, care has to be taken, when analyzing metabolites with high turnover rates due to the long time needed for sampling, so that fast filtration and centrifugation are mostly applied for amino acids, but not for intermediary metabolites of the central carbon metabolism (Bolten et al. 2007).

To date, the most appropriate method to sample sensible compounds is differential analysis (Taymaz-Nikerel et al. 2009). Shortly, two parallel samples are needed. One sample consists of the complete culture broth, while the other one represents the supernatant, obtained via fast filtration. The metabolite concentrations in both samples are then determined independently and the intracellular levels are computed by subtraction. This differential method was successfully implemented to quantify the metabolome of *Aspergillus niger*, *Penicillium chrysogenum* (de Jonge et al. 2012; Lameiras et al. 2015), as well as for *E. coli* in chemostat (Taymaz-Nikerel et al. 2009) and during glucose pulse experiments (De Mey et al. 2010), but has not been extended to other bacteria so far. The correction for metabolites present in the surrounding medium, possible by this approach, is indeed important. Interestingly, almost all metabolites of the central carbon metabolism are present in the supernatant of bacterial cultures in significant amounts (Bolten et al. 2007).

3.3.2 Metabolite Extraction

To date, there are different extraction protocols available. The basic principles used in these protocols to disrupt cells and extract metabolites are extreme changes in temperature (Gonzalez et al. 1997) and pH (Schneider et al. 2002), the application of mechanical force (Pette & Reichmann 1982) and the use of organic solvents (Bligh & Dyer 1959), respectively. However, it appears difficult to identify a particularly beneficial approach out of these possibilities. Extraction with cold methanol and freeze-thaw cycles lead, compared to most other methods, to stable results for most metabolites (Villas-Bôas et al. 2005; Winder et al. 2008; Canelas et al. 2009). Sugars and sugar-phosphates, however, are hardly soluble in pure methanol, while aqueous methanol combined with thawing processes might not fully stop enzymatic activities, thus, leading to potentially falsified concentrations. As an alternative acidic acetonitrile-methanol extraction was used for *E. coli* (Rabinowitz & Kimball 2007) and *S. cerevisiae* (Canelas et al. 2009). For *E. coli* acidic acetonitrile-methanol extraction resulted in higher yields than extraction with methanol water mixtures. Canelas et al. (2009), on the contrary, reported limited efficiency to extract metabolites of higher molecular weight, as well as polar and phosphorylated compounds from *S. cerevisiae*. These contrasting results can be explained by the different properties of the cell walls of *E. coli* and *S. cerevisiae*. The milder conditions of the acidic acetonitrile-methanol protocol lead to an incomplete extraction from eukaryotes and in particular from their cellular compartments, such as mitochondria. The extraction with methanol-chloroform solutions utilizes the denaturation of proteins by chloroform to guarantee the inactivation of enzymes (de Koning & van Dam 1992; Winder et al. 2008). It shows good reproducibility and is especially suitable for sugar phosphates, but not for lipids, because these are retained in the chloroform phase together with the cell pellet (Villas-Bôas et al. 2005; Winder et al. 2008). Extraction with acidic or alkaline solutions, such as perchloric acid or potassium hydroxide can only be used for metabolites that are stable at these extreme pH-values. Villas-Bôas et al. (2005) showed that in perchloric acid only peptides, sugars and sugar alcohols remain stable, whereas amino acids, fatty acids and sugar alcohols are stable in the alkaline extractions. Moreover, such extraction methods have only a rather low reproducibility (Gonzalez et al. 1997; Winder et al. 2008). Furthermore the high salt concentration leads to ion suppression effects during the mass spectrometry (Winder et al. 2008).

A systematic comparison of different extraction methods investigated the reproducibility and the extraction efficiency for 44 metabolites from yeast (Canelas et al. 2009). Extraction with hot water, boiling aqueous ethanol and cold chloroform-methanol, respectively, provided similar metabolite quantities at good reproducibility. In contrast, extraction by freeze-thaw cycles and by acetic-acetonitrile-methanol resulted in lower metabolite abundance and poor reproducibility. Taken together, it can be stated that there is not a single method available to extract the complete metabolome at once, due to the different properties of metabolites. For each particular question, the extraction method has to be chosen and validated individually. At present the most suitable extraction protocol is simple boiling of cells in water or in aqueous ethanol (Villas-Bôas et al. 2005; Winder et al. 2008). Both methods show a high reproducibility, are easy to realize and fast, thus being an excellent starting point for metabolite extraction from bacteria.

3.4 Fundamentals of Metabolism

3.4.1 Central Carbon Metabolic Pathways

The central carbon metabolism is a highly conserved set of amphibolic pathways that occur almost ubiquitous in nature. Core routes of this pathway network are the Embden-Meyerhof-Parnas (EMP) pathway, the pentose phosphate (PP) pathway, the Entner-Doudoroff (ED) pathway and the tricarboxylic (TCA) acid cycle (Figure 3.8). These pathways generate energy, in form of ATP and GTP, redox equivalents, such as NADH, NADPH and FADH, as well as precursor metabolites for biosynthesis. Most of the reactions catalyzed by the enzymes are reversible, thereby, enabling the flexible funneling of different nutrients from varying entry points. To which extent a certain pathway is used, depends on the available carbon source, the needs of the organism and the enzymes that are encoded in the genome.

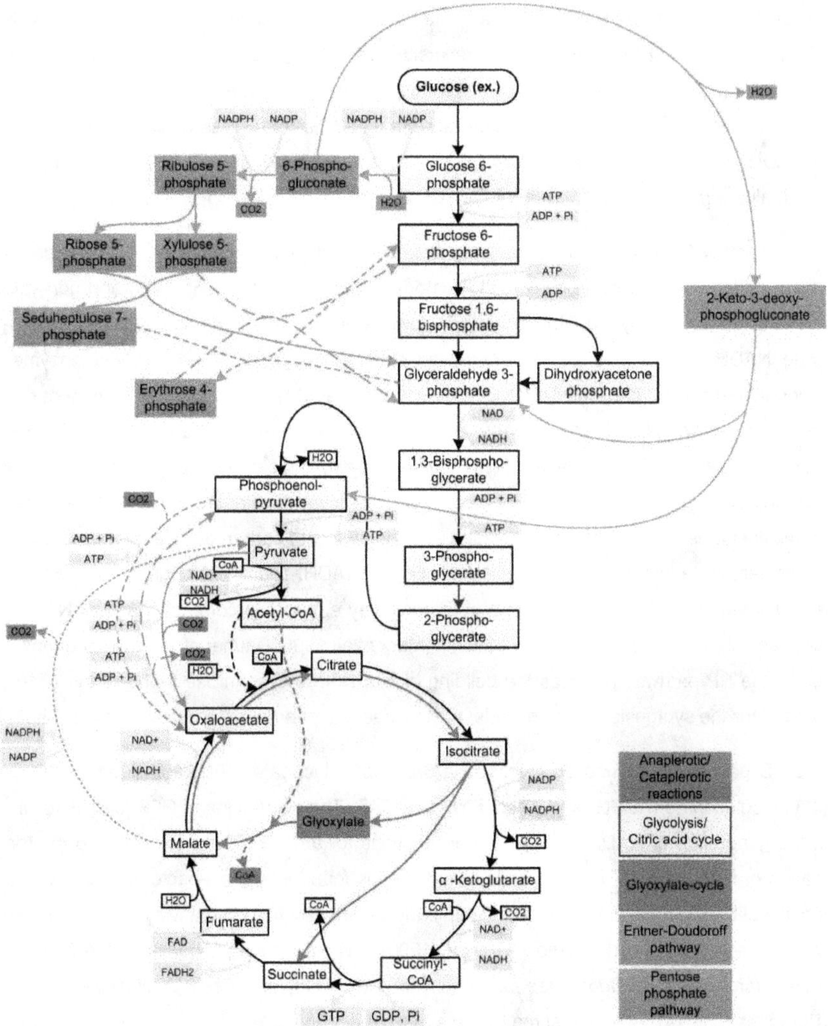

Figure 3.8: Overview of major reactions of the central carbon metabolism, consisting of glycolysis and tricarboxylic acid cycle (white), pentose phosphate pathway (green), Entner-Doudoroff pathway (blue), glyoxylate shunt (red) and other anaplerotic reactions (violet). Production and consumption of energy metabolites and of redox equivalents are shown in yellow.

Many microorganisms assimilate glucose via the phosphotransferase system which transfers a phosphoryl group from phosphoenolpyruvate (PEP) onto the sugar, resulting in glucose 6-phosphate (G6P) (Kotrba et al. 2001). Alternatively, G6P can be acquired by the use of glucose transporter systems and hexokinase activity (Cirillo 1961). G6P is metabolized via different glycolytic routes, i.e. the EMP, the PP and the ED pathway, which all converge at the pyruvate (PYR) node. In the EMP pathway, the irreversible phosphorylation of glucose and fructose 6-phosphate (F6P) by hexokinase and phosphofructokinase, respectively, consumes two molecules of ATP. The lower glycolysis generates four molecules of ATP as well as two molecules of NADH. In addition, the EMP pathway provides 3-phosphoglycerate (3PG), PEP and PYR, which all are metabolic precursors for the synthesis of amino acids, lipids and fatty acids.

The PP pathway consists of an oxidative and a non-oxidative part, respectively. The reactions of the oxidative PP pathway convert G6P into ribulose 5-phosphate (RIBU5P) and generate reducing power. Mostly, NADPH is formed. However, specific bacteria can alternatively generate NADH due to isoenzymes or relaxed substrate specific of the involved enzymes (Tonouchi et al. 2003; Tännler et al. 2008; Klingner et al. 2015). In the non-oxidative part of the PP pathway, RIBU5P is converted into ribose 5-phosphate (R5P) and xylulose 5-phosphate (X5P), respectively. A rearrangement of carbon two and carbon three units further forms F6P and GAP, hereby linking the EMP and the PP pathway. The entire PP pathway stoichiometrically forms 5 molecules of pyruvate and 3 molecules of carbon dioxide out of 3 glucose molecules, while generating 8 ATP, 6 NADPH and 5 NADH. The flux through the PP pathway is highly versatile. The enhanced use of the pathway e.g. to overcome shortage of NADPH has been shown for several microorganisms (Marx et al. 1996; Dauner et al. 2001). For anabolism, the PP pathway provides the building blocks R5P and erythrose 4-phosphate (E4P), needed for the synthesis of amino acids, nucleotides and energy metabolites, respectively.

The ED pathway decomposes glucose via 6-phosphogluconate (6PG) and 2-keto-3-deoxy-phosphogluconate (KDPG) and yields PYR and GAP. The conversion of G6P to 6PG results in the generation of NADPH, while further reactions towards GAP and PYR lead to the formation of NADH and ATP. An active ED pathway is found in *Archea*-bacteria, which lack a functional EMP pathway, thus, indicating an early evolution of the ED pathway as the first form of glucose degradation (Romano & Conway 1996). It is also found in bacteria of the *Pseudomonas* family, like *Pseudomonas putida* (Blank et al. 2008) and *Pseudomonas aeruginosa* (Berger et al. 2014). Although some bacteria, such as *E. coli*, contain all enzymes required for the ED pathway, they favor the EMP pathway to gain energy and precursor metabolites from glucose, due to its higher efficiency in energy generation (Wittmann et al. 2007). In contrast, the ED pathway seems to be preferred under conditions of elevated stress (Klingner et al. 2015).

The TCA cycle can be found throughout almost all bacterial species, because of its importance as starting and end point for diverse anabolic and catabolic reactions. On the one hand, it provides reduction equivalents by the complete oxidation of PYR into three molecules of carbon dioxide. On the other hand, it supplies the building blocks α-ketoglutarate (α-KG) and oxaloacetate (OA). Overall, the TCA cycle decomposes PYR to carbon dioxide while producing two molecules of NADH and one molecule of NADPH, $FADH_2$ and GTP each. Similar to the PP pathway, certain flexibility in coenzyme specificity of isocitrate dehydrogenase and succinyl-coenzyme A synthetase has been reported for specific microorganisms (Kapatral et al. 2000; Banerjee et al. 2005). Furthermore, the cycle provides the amino acid building blocks α-KG and OA. In addition, anaplerotic reactions replenish the TCA cycle. The phosphoenolpyruvate-oxaloacetate-pyruvate-node functions as a link between anabolic and catabolic reactions and is responsible for adapting the distribution of PYR, OA and PEP to fit cellular needs (Sauer & Eikmanns 2005). Enzymes involved are PYR carboxylase, PEP carboxylase, PEP carboxykinase and malic enzyme, whereby the presence of each of these enzymes differs strongly among bacteria. The glyoxylate (GA) shunt enables microorganisms to grow on carbon two substrates (Kornberg 1966). It also refills the TCA cycle via formation of succinate (SUC) and malate (MAL).

3.4.2 Regulation of Metabolism by the Intracellular Energy Level

As described, the main purpose of central carbon metabolism is the supply of sufficient amounts of energy, reducing power and precursors at the very heart of cellular functioning. It has turned out, that this highly interconnected network is strongly controlled by the energy level of the cell (Atkinson & Walton 1967). Key enzymes of the pathways, which are either participating in biosynthesis, energy storage and degradation of energy rich compounds, respectively, are controlled by the ratio of energy rich adenylates to the sum of all adenylates (Atkinson 1968). To describe this important mechanism quantitatively, the adenylate energy charge (AEC) was introduced as fundamental parameter (Atkinson and Walton 1967). It is defined as follows (Equation 3.1).

$$AEC = \frac{[ATP] + 0.5 \cdot [ADP]}{[AMP] + [ADP] + [ATP]} \qquad \text{(Equation 3.1)}$$

The AEC can vary between 0 and 1, depending on the ratio between AMP, ADP and ATP. It was initially postulated that enzymes are not only controlled by the availability of their substrates, but also by the overall energy charge available to the cell, hereby, generating an evolutionary advantage. Exemplified for the activity of the ATP-consuming citrate cleavage enzyme, which plays a major role in the synthesis of fatty acids, a strong dependence on the AEC could be shown. The higher the AEC gets, the higher the reaction rate of the enzyme becomes (Atkinson & Walton 1967). Likewise, ATP-consuming phosphofructokinase is controlled by the energy level. Here, reaction rate drastically decreases with increasing AEC values (Ramaiah et al. 1964). The behavior of these two enzymes is typical for many enzymes involved in the energy metabolism as further measurements of the reaction rate of ATP consuming or regenerating enzymes showed (Shen et al. 1968; Klungsoyr et al. 1968). It was deduced that the intersection of the curves for rates of consuming and regenerating enzymes represents a metabolic steady state (Figure 3.9). A change in the AEC level leads to downregulation of inducing enzymes, while upregulating the counteracting enzymes at the same time, thus buffering the AEC at a value of around 0.85 (Atkinson 1968).

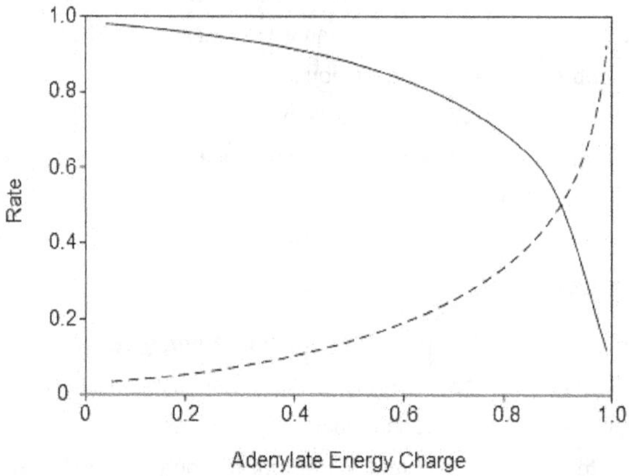

Figure 3.9: Generalized response of ATP-regenerating (solid line) and ATP-consuming (dashed line) enzymes. Figure adapted from Atkinson (1968).

Exceptions might be given in extreme case, e.g. if the balance between ATP utilization and regeneration becomes distorted. It is for example possible to lower the ATP generation rate of the respiratory chain by the addition of the protonophore dinitrophenol. Dinitrophenol acts as a shuttle for protons to cross the biological membranes, thus, leading to a collapse of the proton motive force used by the cell to generate ATP. To compensate the resulting decrease of the ATP-level, the microorganism raises the velocity of other ATP generating pathways like the glycolysis (Dietzler et al. 1979). If this is not possible due to other limitations, the AEC will decrease (Williams & Weiss 1978).

Chapman et al. (1971) examined the shifting of the AEC in *E. coli* depending on the change of the cultivation conditions. The AEC decreased almost instantaneously to a value of 0.5 as a reaction to starvation, but relaxed back to a value of 0.85, immediately after replenishment of the carbon source. In parallel, the overall adenylate concentration decreased in phases of starvation. Obviously, cells remained viable, but probably rather dormant, during periods of starvation. More recently, AEC has been used to validate quantitative metabolome data

(Bolten et al. 2007). Here, AEC values between 0.7-0.95 were considered as physiologically meaningful.

3.4.3 Thermodynamic Constraints

Apart from regulatory constraints of the adenylate energy charge, the second law of thermodynamics can be applied as an even more fundamental approach to evaluate metabolomic datasets. Thermodynamic constraints rely on a physical principle and have to be obeyed, irrespective of the surrounding conditions. Thermodynamic constraints were identified as one of the major driving forces in the evolution of metabolic pathways (Weber et al. 1989; Meléndez-Hevia et al. 1997). Moreover, thermodynamic considerations enable a link between physical laws and complex systems of enzymatic reactions, thus allowing the evaluation of pathway feasibility, based on thermodynamics and measured metabolite concentrations (Cornish-Bowden 1981; Mavrovouniotis 1993).

Shortly, the second law of thermodynamics states that metabolic conversions can only commence if entropy is produced in the process. Thermodynamic feasibility of metabolic reactions is evaluated by changes in the Gibbs energies, which is defined as the difference in Gibbs energies stored in substrates and products of the conversion, namely the Gibbs formation energy (ΔG) (Gibbs 1873). Thus, a reaction will only proceed if ΔG becomes negative (Kümmel et al. 2006). As soon as the Gibbs formation energy becomes positive, the reaction will commence in the backward direction, while a value of zero equals a state of equilibrium. The Gibbs energy of formation is connected to the law of mass action (Equation 3.2), with R being the ideal gas constant, T the ambient temperature and $\Delta G^{\circ\prime}$ the standardized free Gibbs energy (Cornish-Bowden 1981).

$$\Delta G = \Delta G^{\circ\prime} + RTln\frac{\Pi[P]}{\Pi[E]} \qquad \text{(Equation 3.2)}$$

Utilization of the standardized free Gibbs energy is an effective way to consider the impact of ionic strength and pH of the surrounding medium on the actual Gibbs energies of the reaction. Consequently, standard free Gibbs energies are mandatory to evaluate the feasibility of metabolic conversions, as the reactant concentrations can be drawn from the measured metabolic data. Lately, several databases were established, which provide the standard free Gibbs energies of reactions of the central carbon metabolism under physiological conditions, thus enabling the specification of changes in flow direction of metabolic networks, pathways or reactions (Li et al. 2011; Flamholz et al. 2012).

Due to their importance in the proceeding of metabolism, thermodynamic constraints have been applied in the elucidation of several scientific questions. Cornish-Bowden (1981) used such constraints to elucidate, why certain enzymes of the glycolytic pathway function as a bottleneck at differing concentrations of the involved reactants. With increasing computational power, the applications of thermodynamics recently became more complex. Nowadays, thermodynamic constraints are applied to enhance metabolic flux analysis (Henry et al. 2007), to evaluate quantitative metabolome data via computation (anNET) (Zamboni et al. 2008), to assign reaction directionality (Fleming et al. 2009), to evaluate the impact of fluctuations in pH and ionic strength (Vojinović & von Stockar 2009) or to study the regulation of metabolic networks (Garg et al. 2010). One of the youngest application fields of thermodynamic considerations is the improvement of existing pathways (Ducat & Silver 2012) or even the design of synthetic pathways to fit industrial demands (Bar-Even et al. 2010). Hence, thermodynamic considerations have proven to be a powerful tool for the evaluation and elucidation of metabolic datasets. Due to its wide area of applications and its robustness, thermodynamic analysis has become inseparable from metabolomics (Martínez & Nielsen 2014).

3.5 Pathway repertoire of microorganisms

To date, a broad spectrum of microorganisms is heavily researched, engineered and utilized in the biotechnological industry due to their enormous potential in synthesis of high value or bulk chemicals. The variety among the carbon core metabolism of these microorganisms is large, as can be seen in Figure 3.10 (Becker & Wittmann 2015).

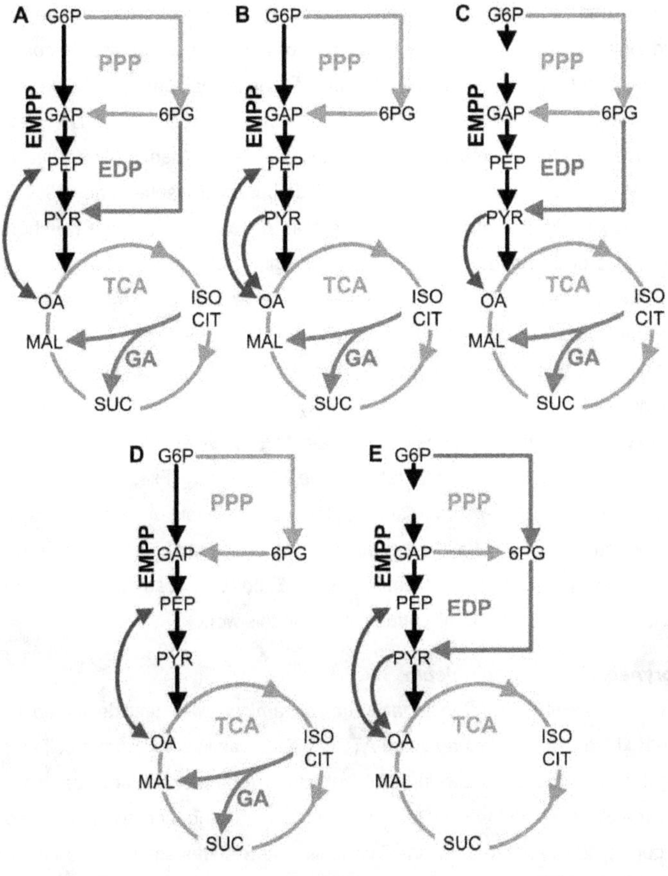

Figure 3.10: Comparative overview depicting the pathway configuration of the central carbon metabolism of various production strains, which were investigated in the course of the present work. (A) *Escherichia coli* and *Yersinia pseudotuberculosis*, (B) *Corynebacterium glutamicum* and *Bacillus megaterium*, (C) *Pseudomonas putida*, (D) *Rhodopseudomonas palustris* and (E) *Dinoroseobacter shibae*

3.5.1 *Escherichia coli*

Escherichia coli is a Gram-negative, rod-shaped, non-sporulating and facultative anaerobic bacterium that was first described by the pediatrician Escherich in 1885 and can be found in the lower intestine of warm-blooded organisms (Shulman et al. 2007). Since its discovery, it has advanced to one of the best researched prokaryotic model organisms. Today, it is heavily utilized in the biotechnological industry and in microbiological research because of the amenability to manipulation and culture (Huang et al. 2012). These characteristics have led to the discovery of biological principles, such as bacterial conjugation (Lederberg & Tatum 1946), topography of gene structure (Benzer 1961) and the foundation of modern biotechnology by Cohen and Boyer (Russo 2003). The production of human insulin using *E. coli* as a host (Williams et al. 1982) was then the first application of genetic engineering in the pharmaceutical industry. The sequencing of the complete genome of strain *E. coli* K12 in 1997 (Blattner et al. 1997) enabled further developments in the production of recombinant proteins, like the correct folding of proteins together with formation of disulfide bonds (Bessette et al. 1999) and the N-linked glycosylation (Wacker et al. 2002). Since then, *E. coli* has become a workhorse of biotechnology and is being used for the production of bulk chemicals as well as highly specific pharmaceuticals (Rodrigues et al. 2014; Becker & Wittmann 2015). *E. coli* itself can grow on different carbon sources and has a highly versatile carbon metabolism (Figure 3.10 A). When growing on glucose, *E. coli* utilizes the EMP-, PP-, and the TCA pathway along with the glyoxylate-shunt and in a minor manner the ED pathway (Rui et al. 2010) to generate energy and metabolic precursors. In the presence of gluconate or gluconic acid the flux distribution shifts from the EMP pathway to an enforced usage of the ED pathway (Fliege et al. 1992). Furthermore, *E. coli* is one of the few microorganisms, for which datasets of the intracellular levels of the central carbon metabolism have been published (De Mey et al. 2010; Hoque et al. 2011). Hence and due to its function as a model organism, *E. coli* was used during the development and validation process of the quantification method in this work.

3.5.2 *Corynebacterium glutamicum*

The Gram-positive bacterium *Corynebacterium glutamicum* was first described by (Udaka 1960). The wild type strain *C. glutamicum* ATCC 13032 was isolated from a soil sample from the elephant cage in the Ueno Zoo in 1957, but can also be found in sewage, manure or on fruits and vegetables. *C. glutamicum* is an aerobic, fast-growing, non-sporulating, non-motile, non-pathogenic, rod-like microorganism. Due to its GRAS-status and based on metabolic engineering, it has been used extensively for the production of amino acids, which are required as food additives for human and animal nutrition as well as pharmaceutical products (Leuchtenberger et al. 2005). The ability to assimilate raw sugars, molasses or lactate from fermented green cuttings (Becker et al. 2005; Andersen & Kiel 2000) together with a completely sequenced genome have made it extremely valuable for biotechnological production. Once discovered as natural glutamate producer, *C. glutamicum* is nowadays utilized for industrial production of a wide variety of amino acids, vitamins and nucleotides (Vertes et al. 2005; Becker & Wittmann 2012a). For growth on glucose as carbon source, *C. glutamicum* uses the EMP- , the PP pathway and the TCA cycle (Marx et al. 1996) (Figure 3.10 B). Activity of the ED pathway could neither be shown for growth on glucose nor gluconate as carbon source (Vallino & Stephanopoulos 1994). As consequence of its industrial relevance, *C. glutamicum* is a well investigated production strain. However, only few metabolite datasets of central carbon metabolism are available so far (Moritz et al. 2002; Bolten et al. 2007). The datasets are hereby limited to certain energy and redox equivalents, while omitting most intermediary metabolites of the core metabolism. In the course of this work, *C. glutamicum* was used to validate the applicability of the quantification method to Gram positive microorganisms.

3.5.3 *Pseudomonas putida*

Pseudomonas putida is a rod-shaped, flagellated, aerobically fast growing, and Gram-negative bacterium which can be easily isolated from soil and water samples as well as plants and animals. The saprotrophic bacterium often functions as a biological control agent for plants, while benefitting from nutrients in their rhizosphere. It induces plant growth and protects the plants from pathogens by secretion of siderophores in exchange (Dupler & Baker 1984; Gravel et al. 2007). Research on *P. putida* is used to elucidate biological principles typical for the *Pseudomonas* clade, thereby, helping to understand pathogenic strains like the closely related *Pseudomonas aeruginosa*, a predominant cause of nosocomial infections. The metabolism of *P. putida* shows high versatility in regard to different nutrients, making it interesting for industrial purposes. On one hand, *P. putida* is capable of degradation of alkyl-benzoates and even polycyclic aromatic compounds like naphthalene and is, therefore, applied for bioremediation of polluted soil (Ronchel et al. 1995; Gomes et al. 2005). On the other hand, *P. putida* is a natural producer of polyhydroxyalkanoates (PHA), a biodegradable plastic. The ease of genomic manipulation in addition to the more recent availability of the complete genome sequence (Nelson et al. 2002) enabled the development of further production strains of biotechnological interest (Poblete-Castro et al. 2012). The central carbon metabolism of *P. putida* differs from that of most bacteria because it lacks the enzyme 6-phosphofructokinase and, consequently, a fully functional EMP pathway (Del Castillo et al. 2007) (Figure 3.10 C). Moreover, *P. putida* uses the PP pathway to generate biomass precursors, but not in a catabolic manner (Ebert et al. 2011). The bacterium instead uses the ED pathway to catabolize glucose and other sugars to generate energy and redox equivalents (Fuhrer et al. 2005). *P. putida* preferably utilizes intermediates of the TCA cycle and therefore suppresses glucose uptake in the presence of TCA cycle metabolites (Wolff et al. 1991). Furthermore, it is capable of degradation of fatty acid via the glyoxylate shunt (Ebert et al. 2011). So far only a few metabolic datasets have been published on *P. putida*, which cover only parts of the central carbon metabolism (Bolten et al. 2007; van der Werf et al. 2008). In the present work, the core metabolome of *P. putida* was quantified, validated by thermodynamic constraints and subsequently compared to the metabolic data of other ED pathway users.

3.5.4 *Yersinia pseudotuberculosis*

Yersinia pseudotuberculosis can be isolated from mammalians with wild rodent being the historically most prominent host (Isaacson et al. 1983). The facultative anaerobe is a Gram-negative, rod-shaped, non-sporulating and motile pathogen (Zurek et al. 2001). The genome of *Y. pseudotubercolosis* was sequenced by Chain et al. (2004) and compared to the genome of *Yersinia pestis*, the bacteria responsible for the bubonic plague. The comparison showed a matching rate of 75 %, indicating a genetic relationship of *Y. pestis* and *Y. pseudotuberculosis* despite drastic differences in pathogenicity and transmission. Motility of the bacterium is regulated by a quorum-sensing system, allowing it to estimate its concentration in the close vicinity and to react accordingly (Atkinson et al. 1999). An infection with *Y. pseudotuberculosis* often results in diarrhea and abdominal pain combined with strong proinflammatory reactions in the infected tissues (Jalava et al. 2006; Logsdon & Mecsas 2006). The bacterium is used as a model organism to study the infection process during which the bacteria survive in different environments. The effect of changes in substrate availability on the virulence of *Y. pseudotuberculosis* has recently been clarified (Bücker et al. 2014). Under glucose excess, *Y. pseudotuberculosis* employs a combination of glycolysis and the PP pathway in combination with the tricarboxylic acid cycle to generate energy and biomass for maintenance and replication (Bücker et al. 2014) (Figure 3.10 A). In the absence of glucose, the bacterium is capable of expressing isocitrate lyase and malate synthase, thereby, enabling the utilization of two-carbon molecules or fatty acids as a carbon source via the GA-shunt (Hillier & Charnetzky 1981). An active ED pathway has not been reported in the presence of glucose, even though small amounts of the participating enzymes could be shown. In the presence of gluconate, however, the ED pathway is active (Brubaker 1968). It has been shown that the pathogen tends to accumulate large amounts of extracellular PYR (Bücker et al. 2014). The present work was used to elucidate how *Y. pseudotuberculosis* achieves these high concentrations of PYR. This appeared especially interesting, as to date, no metabolic datasets of the central carbon metabolism of the human pathogen *Y. pseudotuberculosis* have been published.

3.5.5 *Bacillus megaterium*

In 1884, Anton de Bary described a Gram-positive soil bacterium and named it, due to its enormous cell size, *Bacillus megaterium* (de Bary 1884). It was this characteristic that made *B. megaterium* the perfect candidate for investigations of bacterial structure, membrane properties, protein localization and sporulation principles (Foerster & Foster 1966; Christie et al. 2010), thus, *B. megaterium* became a model organism long before *Bacillus subtilis*. The rod-like, non-pathogenic, motile bacterium is aerobically fast growing and capable of sporulation to endure harmful conditions (Eppinger et al. 2011). *B. megaterium* is ubiquitous in the environment and tolerating extreme conditions as well as using numerous different carbon sources to sustain its growth (Vary et al. 2007; Lüders et al. 2011; Khan 2011). These features in combination with the absence of endotoxins and alkaline proteases made the bacterium a promising target for the development of biotechnological production processes (Bunk et al. 2010). It has proven to be easily accessible for genetic modification by different techniques. Thus, it has been used for the production of enzymes, such as amylase and penicillin amidase and for aerobic and anaerobic vitamin B_{12} generation (Vary et al. 2007; Korneli et al. 2013). The genome sequence of two important *B. megaterium* strains were recently published (Eppinger et al. (2011), thus facilitating a deeper understanding of its intracellular fluxes and their regulation. Application of glucose as carbon source results in energy generation by a combination of the EMP and PP pathway followed by the TCA cycle (Wang et al. 2005) (Figure 3.10 B). Additionally, *B. megaterium* carries genes encoding for the glyoxylate-shunt, allowing growth on 2-carbon compounds (Eppinger et al. 2011). An active ED pathway could neither be shown for glucose- nor gluconate-dependent growth (Otani et al. 1986; Otani et al. 1987). *B. megaterium* is heavily utilized for industrial production of proteins. During such production processes the microorganism is constantly stressed, which might influence productivity negatively. Thus, quantitative metabolomics was applied to identify the consequences of stress on the central metabolism. Despite its industrial relevance, no metabolic datasets of the core metabolism of *B. megaterium* are available at present.

3.5.6 *Rhodopseudomonas palustris*

One of the metabolically most versatile microorganisms described so far, is the Gram-negative, purple non-sulfur bacterium *Rhodopseudomonas palustris* (Larimer et al. 2004). The rod-shaped, motile, non-pathogenic and facultative anaerobe prokaryote is widely spread as the isolation from different aquatic habitats, coastal sediments and even swine waste lagoons demonstrates (Larimer et al. 2004). The bacterium utilizes energy, generated from light or organic compounds, to maintain growth. It is capable to assimilate carbon from organic and inorganic compounds as well as elemental nitrogen (Klemme et al. 1980). Additionally, the versatile carbon metabolism of *R. palustris* enables the bacterium to degrade sugars, acetates, but also more complex compounds like lignin and benzoate (Hädicke et al. 2011). The genome of *R. palustris* was recently sequenced (Larimer et al. 2004), thereby allowing its use as a model organism for the response of microorganisms to environmental changes. The bacterium is used to elucidate the working principles of bacterial photosynthesis (Roszak et al. 2003) and to generate electricity by biological fuel cells (Xing et al. 2008). Moreover, the capability of benzoate degradation in combination with the production of hydrogen makes it a promising candidate for bioremediation and generation of clean energy (Fißler et al. 1995). Another possible application field of purple non-sulfur bacteria is their usage as biological control agents for aquacultures as a sustainable replacement of antibiotics and vitamins (Kim & Lee 2000).

R. palustris favorably uses succinate as carbon source for its growth by assimilation via the oxidative TCA cycle, but is also capable of energy and biomass generation from sugars with glycolysis, PP pathway and glyoxylate shunt (Eley et al. 1979; Imai et al. 1984; Hädicke et al. 2011) (Figure 3.10 D). Under anaerobic conditions, *R. palustris* assimilates carbon dioxide by the Calvin-Benson-Bessham (CBB) cycle (Joshi et al. 2009). Probably due to the fact, that industrial relevance of purple non-sulfur bacteria is still low, no datasets of metabolite levels are available yet. In the course of this work, metabolite levels of the photoheterotrophically growing *R. palustris* were compared to metabolite levels of microorganisms employing gluconeogenesis.

3.5.7 *Dinoroseobacter shibae*

Dinoroseobacter shibae is a Gram-negative, marine, small oval rod-shaped bacterium. It belongs to the group of aerobic anoxygenic phototrophs (AAP) and develops a dark red pigmentation if grown in darkness. It can either grow planktonic or form a biofilm on phytoplankton, as its first isolation by washing of the dinoflagellate *Prorocentrum lima* proves (Biebl et al. 2005). The motile, non-pathogenic bacterium is capable of aerobic and anaerobic photosynthesis, which is used to generate additional energy under heterotrophic conditions (Wagner-Döbler et al. 2010). Moreover, *D. shibae* is adapted to light-dark cycles as shown by analysis of its transcriptional dynamics and long term starvation experiments (Tomasch et al. 2011; Soora & Cypionka 2013). These features, together with the ability to produce storage compounds during excess conditions show the perfect adaption of *D. shibae* to its marine habitat (Rex et al. 2013). The bacterium, furthermore, resorts to arginine fermentation or nitrate reduction to cope with anaerobic conditions (Laass et al. 2014; Ebert et al. 2013). The available genome sequence reveals additional metabolic traits, like degradation of aromatic compounds and the oxidation of carbon monoxide (Wagner-Döbler et al. 2010). If living in a symbiotic relationship with microalgae, *D. shibae* produces the vitamins B1 and B12 in exchange for organic acids, which are converted into energy and biomass building blocks by the TCA cycle (Wagner-Döbler et al. 2010). In addition, labeling studies showed, that *D. shibae*, while having the ED- and the EMP pathway at its disposal, exclusively degrades glucose via the ED pathway (Fürch et al. 2009) (Figure 3.10 E). The reductive PP pathway is solely used for the generation of metabolic precursors, as the 6-phosphogluconate dehydrogenase is not annotated (Wagner-Döbler et al. 2010). The assimilation of acetate via the GA cycle is also not possible, as the isocitrate lyase is not encoded in its genome. *D. shibae* was only recently isolated, thus no metabolic data covering the central carbon metabolism is available so far. However, the unconventional pathway usage, the scarce habitat and its broad substrate specificity made it an interesting candidate for metabolite quantitation. Here the influence of a change of substrates on the intracellular metabolite pools could be elucidated.

4 Material and Methods

4.1 Strains

Different wild type microorganisms, used in the present work, were obtained from public strain collections, such as the American Type Culture Collection (ATCC) and the German Collection of Microorganisms and Cell Cultures (DSMZ), respectively (Table 4.1). In addition, specifically designed mutants of *Y. pseudotuberculosis* were derived from a previous work (Bücker et al. 2014) (Table 4.2).

Table 4.1: Wild type bacterial strains used in the present work

Strain	Source
Escherichia coli K12 DSM 2670	DSMZ
Corynebacterium glutamicum ATCC 13032	ATCC
Pseudomonas putida KT2440 DSM 6125	DSMZ
Yersinia pseudotuberculosis YPIII DSM 8992	DSMZ
Bacillus megaterium DSM 319	DSMZ
Rhodopseudomonas palustris DSM 123	DSMZ
Dinoroseobacter shibae DFL-12 DSM16493	DSMZ

For long term cryoconservation, strains were grown in their corresponding complex medium and under permanent light if necessary (*D. shibae* and *R. palustris*). Cell suspension (500 µL) was harvested in mid exponential phase, transferred into stock culture tubes and mixed with 500 µL of a 60 % glycerol solution. After thorough mixing, the tubes were transferred into liquid nitrogen, thereby shock freezing the culture. Cryo-cultures were subsequently stored at -80 °C.

Table 4.2: Mutants of *Y. pseudotuberculosis* used in the present work

Strain	Modifications	Source
Y. pseudotuberculosis YPIII Δ*pykF*	Deletion of pyruvate kinase F, Kanamycin resistance	Bücker et al., 2014
Y. pseudotuberculosis YPIII Δ*ptsN*	Deletion of the nitrogen regulatory protein PtsN, Kanamycin resistance	Bücker et al., 2014
Y. pseudotuberculosis YPIII Δ*arcA*	Deletion of the global anaerobiosis response regulator, Kanamycin resistance	Bücker et al., 2014
Y. pseudotuberculosis YPIII Δ*pdhR*	Deletion of the pyruvate dehydrogenase complex regulator PdhR, Kanamycin resistance	Bücker et al., 2014

4.2 Chemicals

Tryptone and yeast extract were purchased from BD Biosciences (Heidelberg, Germany). HAM F-12 Nutrient mixture was obtained from Invitrogen (Carlsbad, California, United States) and liquid DMEM was obtained from Biochrom (Berlin, Germany). Marine broth supplement (MB2216) was purchased from Becton Dickinson (Franklin Lakes, New Jersey, United States). Labeled $U^{13}C$-glucose was purchased from Eurisotop (Saarbrücken, Germany). All other chemicals were obtained from Sigma-Aldrich (Steinheim, Germany), Merck (Darmstadt, Germany) and Fluka (Buchs, Switzerland). Usually, chemicals were used at analytical grade. Metabolites and solvents, used for LC-MS/MS tuning, calibration and sample analysis, were obtained in the highest purity available. Ultrapure water (resistivity < 18.2 MΩ at 25 °C) was obtained by utilization of Milli-Q integral water purification system (Millipore, Merck KGaA, Darmstadt, Germany).

4.3 Growth Media

4.3.1 Complex Media

For cultivation of first pre-cultures of *E. coli*, *B. megaterium*, *C. glutamicum* and *P. putida*, respectively, LB-medium in baffled shake flasks was used (Table 4.3).

Table 4.3: Composition of the LB-complex medium used for cultivation of *E. coli, B. subtilis, B. megaterium, C. glutamicum* and *P. putida*

Compound	Quantity
Tryptone	10 g
Yeast extract	5 g
NaCl	10 g

ad 1 L aqua dest.; autoclaved (121 °C, 20 min)

First pre-cultures of *Y. pseudotuberculosis* were grown in sterile-filtered 1:1 mixture of HAM's F-12 nutrient mixture and liquid DMEM. Pre-cultures of *D. shibae* were grown in marine broth complex medium (Table 4.4).

Table 4.4: Composition of the complex medium used for cultivation of *D. shibae*

Compound	Quantity
Difco Marine Broth (MB 2216)	37.4 g

ad 1 L aqua dest.; autoclaved (121 °C, 20 min)

For preparation of agar plates, 1.5 % (w/v) of agar Difco (Detroit, Michigan, United States) was added to the corresponding medium, prior to sterilization.

4.3.2 Minimal Media

For second pre-culture and main culture, different minimal media were used. *E. coli* and *C. glutamicum* were grown in glucose minimal medium (Becker et al. 2010) (Table 4.5).

Table 4.5: Composition of the minimal medium used for cultivation of *C. glutamicum* and *E. coli*

Compound	Quantity
Glucose	10 g
NaCl	1 g
$CaCl_2$	55 mg
$MgSO_4 \cdot 7\,H_2O$	0.2 g
$(NH_4)_2SO_4$	15 g
K_2HPO_4	34.8 g
KH_2PO_4	8.5 g
$FeSO_4 \cdot 7\,H_2O$	20 mg
$FeCl_3 \cdot 6\,H_2O$	2 mg
$MnSO_4 \cdot H_2O$	2 mg
$ZnSO_4 \cdot H_2O$	0.5 mg
$CuCl_2 \cdot 2\,H_2O$	0.2 mg
$Na_2B_4O_7 \cdot 10\,H_2O$	0.2 mg
$Na_2MoO_4 \cdot 4\,H_2O$	0.1 mg
Biotin	0.5 mg
Thiamin · HCl	1 mg
Panthothenic acid Ca-Salt	1 mg
3,4-Dihydroxybenzoic acid	30 mg

ad 1 L aqua dest.; sterile-filtered; pH 7,2

In selected experiments, the glucose concentration was varied as specified below. Table 4.6 shows the minimal medium used for cultivation of *E. coli* according to Chapman et al. (1971).

Table 4.6: Composition of the minimal medium for cultivation of *E. coli* during starvation experiments

Compound	Quantity
Glucose	1 g
$(NH_4)_2SO_4$	2 g
K_2HPO_4	5 g
KH_2PO_4	13 g
$MgCl_2$	0.2 g

ad 1 L aqua dest.; autoclaved (121 °C, 20 min); pH 7,2

Table 4.7 shows the composition of M9 minimal medium, used for the cultivation of *B. megaterium*. During cultivation under osmotic stress, the NaCl concentration was increased from 0.5 g/L to 35.06 g/l and 70.13 g/L to achieve molarities of 0.6 M and 1.2 M, respectively.

Table 4.7: Composition of the minimal medium used for cultivation of *B. megaterium*

Compound	Quantity
Glucose	10 g
$Na_2HPO_4 \cdot 2\ H_2O$	8.5 g
KH_2PO_4	3 g
NH_4Cl	1 g
NaCl	0.5 g
$MgSO_4 \cdot 7\ H_2O$	246 mg
3,4-Dihydroxybenzoic acid	30 mg
$CaCl_2 \cdot 2\ H_2O$	14.7 mg
$FeCl_3 \cdot 6\ H_2O$	13.5 mg
$ZnCl_2$	1.7 mg
$MnCl_2 \cdot 4\ H_2O$	1 mg
$CoCl_2 \cdot 6\ H_2O$	0.6 mg
$Na_2MoO_4 \cdot 2H_2O$	0.6 mg
$CuCl_2 \cdot 2\ H_2O$	0.43 mg

ad 1 L aqua dest.; autoclaved (121 °C, 20 min); pH 7,2

Cultivation of *P. putida* was performed in modified M9 minimal medium (Poblete Castro 2012). Its composition is depicted in Table 4.8.

Table 4.8: Composition of the minimal medium used for cultivation of *P. putida*

Compound	Quantity
Glucose	10 g
$Na_2HPO_4 \cdot 2\ H_2O$	8.5 g
KH_2PO_4	3 g
$(NH_4)_2SO_4$	4.7 g
NaCl	0.5 g
$MgSO_4 \cdot 7\ H_2O$	0.12 g
$FeSO_4 \cdot 7\ H_2O$	6 mg
$CaCO_3$	2.7 mg
$ZnSO_4 \cdot H_2O$	2 mg
$MnSO_4 \cdot H_2O$	1,16 mg
$CoSO_4 \cdot 7\ H_2O$	0.37 mg
$CuSO_4 \cdot 5\ H_2O$	0.33 mg
H_3BO_3	0.08 mg

ad 1 L aqua dest.; autoclaved (121 °C, 20 min); pH 7,2

Cultivation of *Y. pseudotuberculosis* was carried out in Yersinia Minimal Medium (YMM) (Bücker et al. 2014). The medium composition is shown in Table 4.9.

Table 4.9: Composition of the Yersinia Minimal Medium used for cultivation of *Y. pseudotuberculosis*

Compound	Quantity
Glucose	8 g
KH_2PO_4	13.26 g
K_2HPO_4	6.62 g
$(NH_4)_2SO_4$	5 g
NaCl	0.31 g
$MgSO_4 \cdot 7\ H_2O$	0.2 g
3,4-Dihydroxybenzoic acid	30 mg
$FeCl_3 \cdot 6\ H_2O$	2 mg
$MnSO_4 \cdot H_2O$	2 mg
$ZnSO_4 \cdot 7\ H_2O$	1.3 mg
$FeSO_4 \cdot 7\ H_2O$	0.5 mg
$CuCl_2 \cdot 2\ H_2O$	0.2 mg
$Na_2B_4O_7 \cdot 10\ H_2O$	0.2 mg
$(NH_4)_6Mo_7O_{24} \cdot 4\ H_2O$	0.1 mg

ad 1 L aqua dest.; autoclaved (121 °C, 20 min); pH 7,2

Table 4.10 depicts the composition of modified minimal medium, used to grow *R. palustris* (Lippe 2008). To enable photoheterotrophic growth, the medium had to be stripped from oxygen. To this end, the medium was heated to the boiling point and simultaneously sparged with nitrogen (5 min). Afterwards, the vessel was sealed airtight by a butyl-septum and cooled down to room temperature. Subsequently, vitamins and amino acids were added under sterile conditions.

Table 4.10: Composition of the minimal medium used for cultivation of *R. palustris*

Compound	Quantity
Ethanol	0.5 mL
Na_2-Succinate	1 g
NH_4-Acetate	0.5 g
Ferric(III)-Citrate	5 mg
KH_2PO_4	0.5 g
$MgSO_4 \cdot 7\ H_2O$	0.4 g
NaCl	0.4 g
NH_4Cl	0.4 g
$CaCl_2 \cdot 2\ H_2O$	50 mg
$ZnSO_4 \cdot 7\ H_2O$	0.1 mg
$MnCl_2 \cdot 4\ H_2O$	0.03 mg
H_3BO_3	0.3 mg
$CoCl_2 \cdot 6\ H_2O$	0.2 mg

Compound	Quantity
$CuCl_2 \cdot 2\ H_2O$	0.01 mg
$NiCl_2 \cdot 2\ H_2O$	0.02 mg
$Na_2MoO_4 \cdot 2H_2O$	0.03 mg
Vitamin B_{12}	0.04 mg
Thiamin (B_1)	0.05 mg
Nicotinic acid amide (B_3)	0.05 mg
Biotin (B_7)	0.001 mg
Alanine	60 mg
Aspartic acid	54 mg
Glutamic acid	90 mg
Leucine	54 mg
Lysine	51 mg

ad 1 L aqua dest.; sterile-filtered; pH 7,2

D. shibae was grown in modified artificial seawater medium (Bartsch 2015). The basic medium composition is listed in Table 4.11. As sole carbon source, succinate or glucose, respectively, were added to concentrations of 10 mM.

Table 4.11: Composition of the modified artificial seawater medium used for cultivation of *D. shibae*

Compound	Quantity
KH_2PO_4	0.2 g
NH_4Cl	0.25 g
$NaSO_4$	4 g
$MgCl_2 \cdot 6\ H_2O$	9 g
$CaCl_2 \cdot 2\ H_2O$	0.15 g
$NaHCO_3$	0.19 g
NaCl	20 g
KCl	0.5 g
Triplex III (Na_2EDTA)	10.4 mg
$FeSO_4 \cdot 7\ H_2O$	4.2 mg
H_3BO_3	0.06 mg
$MnCl_2 \cdot 4\ H_2O$	0.2 mg
$CoCl_2 \cdot 6\ H_2O$	0.38 mg
$NiCl_2 \cdot 6\ H_2O$	0.048 mg
$CuCl_2 \cdot 2\ H_2O$	0.004 mg
$ZnSO_4 \cdot 7\ H_2O$	0.288 mg
$Na_2MoO_4 \cdot 2H_2O$	0.072 mg
Biotin	4 mg
Niacin	40 mg
4-Aminobenzoic acid	16 mg

ad 1 L aqua dest.; sterile-filtered; pH 7,2

4.4 Cultivation

4.4.1 Shake Flask Cultivation

E. coli, *C. glutamicum*, *B. megaterium*, *P. putida*, *Y. pseudotuberculosis* and *D. shibae* were maintained on agar plates. First pre-cultures were conducted using baffled shake flasks (100 mL flask volume) containing 10 mL of strain specific complex medium (see above). After 8 h, cells were harvested by centrifugation (10.000 *x g*, 5min) and washed with saline 1.8 % (w/v) NaCl solution to remove residues of complex media. Cells were then used to inoculate a second pre-culture in 25 mL minimal medium using a baffled shake flask (250 mL flask volume). The cells were grown for 12 h, harvested and washed as described previously and finally used as the inoculum of the main culture (25 mL filling volume, 250 mL baffled shake flask).

R. palustris was inoculated from cryo-culture into its corresponding anaerobic minimal medium (50 mL filling volume, airtight sealed 100mL bottles). The shaker was equipped with an illumination unit (3 × 18 W Biolux, Osram, Munich, Germany) to ensure continuous and consistent illumination. A brief summary of the cultivation conditions of all seven strains is given in Table 4.12.

Table 4.12: General cultivation parameters used for cultivation

Strain	Temperature [°C]	Shaking frequency [rpm]	Growth regime
E. coli	30	230	aerobic
C. glutamicum	30	230	aerobic
B. megaterium	37	230	aerobic
P. putida	30	230	aerobic
Y. pseudotuberculosis	25	200	aerobic
D. shibae	30	180	aerobic
R. palustris	25	120	anaerobic, constant light

Main cultures for all experiments were started with an optical density of 0.1 – 0.5 and were performed in triplicate. Cultivations were performed using a rotary shaker (5 cm shaking diameter, Multitron, Infors, Bottmingen, Switzerland). To avoid oxygen limitation the filling volume did not exceed 10 % of the total shake flasks volume.

4.4.2 Fed-batch and Chemostat Cultivations in Bioreactors

Fed-batch and chemostat cultivations were performed in a 1L-bioreactor system with a working volume of 400 mL, kept at a stirring speed of 600 rpm and at an aeration rate of 0.25 vvm (DASGIP, Jülich, Germany). The stirrer speed was adapted to ensure dissolved oxygen levels of > 20 %. Culture settings were controlled at the corresponding temperature (Table 4.12) and pH 7, respectively. The pH was maintained by computer-controlled titration with 1 M NH_4OH and with 1 M HCl, respectively. Process variables were monitored and collected by a process control system (DASGIP Control, DASGIP, Jülich, Germany) Foam formation was prevented by addition of sterile-filtered Ucolub (FRAGOL, Mülheim, Germany).

For fed-batch cultivations, applied during generation of the labeled internal standard with *C. glutamicum* and the investigation of the energy metabolism of *E. coli*, an exponential feed profile was utilized to ensure a constant growth rate. The feed rate was calculated with the help of Equation 4.1 (Yee & Blanch 1992), with μ being the desired growth rate and S_0 being the feed concentration. The yield coefficient had to be determined in preliminary experiments.

$$F = \frac{\mu \times X_0 \times V_0 \times e^{\mu \times t}}{S_0 \times Y_{X/S}} \qquad \text{(Equation 4.1)}$$

Chemostat experiments were conducted at different dilution rates. To achieve dilution rates of $D=0.1\ h^{-1}$, $D=0.2\ h^{-1}$ and $D=0.4\ h^{-1}$, feed medium was pumped into the reactor at pump rates of 40 mL/h, 80 mL/h and 160 mL/h, respectively. Culture broth was removed from the reactor through a steel pipe, which was adjusted to the initial fill level, at a slightly higher pump rate, thereby ensuring constant culture volumes in the reactor. Both reservoirs, the storage tank containing the fresh medium as well as the collection tank for the removed culture broth, were placed on analytical balances, thus enabling a continuous control of pump rates for accurate mass balancing. Samples for quantitative metabolite analysis were taken after five volume changes to guarantee metabolic and isotopic steady state (Zamboni et al. 2009).

4.5 Analytical Techniques

4.5.1 Cell concentration

Cell concentration was measured by optical density (OD) at specific wavelengths (Table 4.13) using a photometer (Libra S11, Biochrom Ltd., Cambridge, England). Samples were diluted on a microbalance (Sartorius, Göttingen, Germany) to values below 0.3 and measured in duplicate against blank medium as reference.

Table 4.13: Wavelengths used for the estimation of cellular growth

Strain	Wavelength λ [nm]
E. coli	660
C. glutamicum	660
B. megaterium	600
P. putida	600
Y. pseudotuberculosis	600
D. shibae	650
R. palustris	660

The cellular dry weight (CDW) was determined after harvesting 10 mL of culture broth by filtration (0.2 µm pore size, RC membrane filters, Sartorius, Göttingen, Germany). Filters were pre-dried at 80 °C for 24 hours and weighed prior to use. After filtration, two washing steps with 0.9 % NaCl and deionized water were applied. Subsequently, filters were dried at 80 °C until constant weight.

For the exact determination of specific intracellular metabolite concentrations and biomass yield coefficients, it was necessary to establish a correlation between optical density and cellular dry weight. From comparative measurements, the optical density was plotted against the corresponding cellular dry weight to calculate the correlation factor. This was done for *E. coli*, *C. glutamicum* and *R. palustris*, as representatively shown for *P. putida* (Figure 4.1).

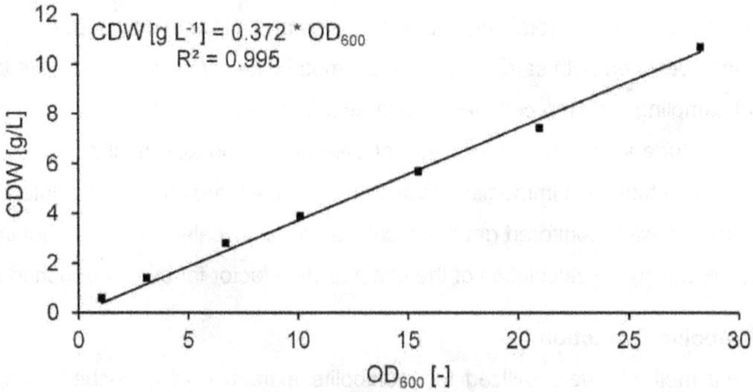

Figure 4.1: Representative correlation of Cellular Dry Weight (CDW) and Optical Density (OD) for *P. putida*

The factors (Table 4.14) were different for each strain, as a consequence of different cell size, shape or substrate. Correlation factors of *B. megaterium*, *Y. pseudotuberculosis* and *D. shibae* were obtained from literature. All measurements were performed with the same photometer that has been used for the present work (Libra S11, Biochrom Ltd., Cambridge, England).

Table 4.14: List of correlation factors used to estimate the actual biomass present in the biological samples taken for metabolite quantification. *D. shibae* exhibits different correlation factors if grown on (a) glucose and (b) succinate as the carbon source.

Strain	f=CDW/OD	Reference
E. coli	0.307	This work
C. glutamicum	0.249	This work
B. megaterium	0.217	Godard (2015)
P. putida	0.372	This work
Y. pseudotuberculosis	0.325	Bücker (2014)
D. shibae (a)	0.383	Bartsch (2015)
D. shibae (b)	0.321	Bartsch (2015)
R. palustris	0.302	This work

4.5.2 Glucose concentration

Concentration of glucose in culture supernatant was measured by a glucose analyzer (YSI 2700 select, YSI Incorporated, Yellow Spring, USA). Prior to measurement, 500 µL of sample were centrifuged (10.000 *x g*, 2 min) to remove cells and cell debris from the broth.

4.6 Metabolomics Workflow

4.6.1 Sampling

The differential method was used to avoid metabolite leakage during the quenching procedure (Taymaz-Nikerel et al. 2009). For this purpose, two samples were taken per sampling point (Figure 3.7). A broth sample was quenched, thus containing metabolites, which were present in the supernatant and in the microorganisms at that time. The second sample was sterile-filtered (0.2 µm pore size, Minisart-plus syringe filter, Sartorius, Göttingen, Germany) into the quenching solution, thereby retaining microorganisms and allowing only supernatant to pass. The difference between both samples equals the metabolite concentrations inside the cells at the time of sampling. To stop cell metabolism as fast as possible, 1 mL sample was directly injected into a tube which contained 5 mL of pre-cooled aqueous methanol (-30 °C, 60 % MeOH (aq.)). The tube was immediately vortexed and afterwards transferred into liquid nitrogen. The process was monitored gravimetrically on a microscale (Sartorius, Göttingen, Germany), thus enabling the calculation of the exact dilution factor for every quenched sample.

4.6.2 Metabolite Extraction

Two different methods were utilized for metabolite extraction of quenched samples. One method used boiling ethanol (75 % (v/v) EtOH (aq.)) as the extraction solvent (Gonzalez et al. 1997), while the second method utilized pre-cooled acidic acetonitrile-ethanol (0.1 M formic acid, 60 % (v/v) acetonitrile, 20 % MeOH (v/v) (aq.)) as metabolite solvent (Rabinowitz & Kimball 2007).

For extraction in boiling ethanol, 0.5 mL of quenched sample was mixed with 100 µL of internal $U^{13}C$-standard and 2.5 mL of the 75 % EtOH (v/v) (aq.) solution. The mixture was incubated at 80 °C for 3 min and afterwards cooled down on ice to minimize degradation.

The acidic acetonitrile-methanol (aq.) solution was cooled down to -20 °C, prior to extraction. Similar to the protocol with boiling ethanol, 2.5 mL extraction solvent was then mixed with 100 µL internal ^{13}C- standard and 0.5 mL of the quenched sample and incubated at -20 °C for 15 min. Afterwards, 50 µL of a 10 % NH_4OH solution was added to neutralize the pH.

After extraction, extracts were centrifuged (5 min, 3000 *x g*) (Heraeus Multifuge 4 KR, Thermo Fisher Scientific Inc, Waltham, United States) to remove remaining cell debris. The supernatant was carefully decanted, frozen at -80 °C and finally lyophilized. The dried extracts were kept at -80 °C until measurement. Prior to measurement, dried extracts were dissolved in 500 µL of eluent A (6 mM tributylamine (aq.) adjusted to pH 6.2 with 6 mM acetic acid). As previously described, all dilution steps were monitored gravimetrically to ensure exact quantification.

4.6.3 Generation of an $U^{13}C$-labeled Internal Standard

For generation of an $U^{13}C$-labeled internal standard, *C. glutamicum* was grown in 0.4 L of minimal medium, which contained 32 g/L of [$^{13}C_6$] glucose as sole carbon source. To minimize the input of ^{12}C carbon, the second pre-culture was also grown on [$^{13}C_6$]-glucose. During the main cultivation in the bioreactor, inlet air was purged through 4M KOH before entering the bioreactor, to remove ambient carbon dioxide (Wu et al. 2005). At an OD_{660nm} of 60, 300 mL of culture broth was directly transferred into a precooled (-40 °C) aqueous MeOH solution (1,5 L; 60 % (v/v)). The quenched culture broth was transferred into 15 mL tubes and centrifuged (2000 *x g*, 5 min, -20 °C). The supernatant was carefully discarded and the cell pellet was extracted (3 min) in 10 mL of boiling aqueous ethanol (75 % (v/v)). The extract was then cooled down. Cell debris was removed by another centrifugation step (2000 *x g*, 5 min, - 20 °C). The supernatant was collected, lyophilized to remove the remaining ethanol and afterwards dissolved in 100 mL of ultrapure water. Aliquots (1 mL each) of the internal standard were prepared and stored at -80 °C.

4.6.4 LC-MS/MS Measurement

All measurements were accomplished with an Agilent 1290 series binary HPLC system (Agilent Technologies, Waldbronn, Germany) coupled with an AB Sciex QTRAP® 5500 linear ion trap mass spectrometer (MS) (AB Sciex, Darmstadt, Germany), equipped with a TurboIonSpray source. Data was acquired and evaluated with the Analyst software (version 1.6, AB Sciex, Darmstadt, Germany). The MS was operated in negative ionization mode and in multiple reaction monitoring (MRM) mode. Single metabolite standards, dissolved in eluent A, were infused at a flow rate of 7 µL min^{-1} to tune compound dependent MS parameters. The MS/MS fragmentation pattern of each analyte was determined and the declustering potential (DP), collision energy (CE) and cell exit potential (CXP) were optimized for maximal intensity. The entrance potential (EP) was fixed at -10 V and the dwell time was set to 5 ms for all transitions. Source dependent MS parameters were set as follows: ion spray voltage (IS) -4500 V, nebulizer gas (GS1) auxiliary gas (GS2), curtain gas (CUR) and collision gas (CAD) were set 60, 60, 35 and medium, respectively. The auxiliary gas temperature was set to 550 °C. The mass spectrometer was run in unit resolution to obtain adequate selectivity and sensitivity.

Table 4.15: Gradient profile applied in the developed LC-MS/MS method. Eluent A: 6mM tributylamine aqueous solution adjusted to pH 6.2 with 6 mM acetic acid, Eluent B: 6mM tributylamine in a mixture of water/acetonitrile 50:50 (v/v) adjusted to pH 6.2 with 6 mM acetic acid.

Step	Total time [min]	Eluent A [vol. %]	Eluent B [vol. %]
1	0.00	95.0	5.0
2	2.00	95.0	5.0
3	22.00	10.0	90.0
4	23.00	95.0	5.0
5	28.00	95.0	5.0

Comparative measurements were done with a previously developed chromatographic separation method (Luo et al. 2007). For optimized chromatic separation, a VisionHT C18 HL 100 mm × 2 mm I.D:, 1.5 µm particles column (Grace, Columbia, MD, United States) at 50 °C was used with eluent A (6 mM aqueous tributylamine solution, adjusted to pH 6.2 with acetic acid) and eluent B (aqueous acetonitrile solution (50% v/v) with 6 mM tributylamine, adjusted to pH 6.2 with acetic acid) The exact gradient profile is shown in Table 4.15. The injection volume was 10 µL. The mobile phase was introduced into the mass spectrometer via the turbo ion spray source at a flow rate of 350 µL min^{-1}.

4.7 Limits of Detection and Quantitation

It is inevitable for successful implementation of analytical methods to determine the limit of detection (LOD) and the limit of quantitation (LOQ). According to their definition in DIN 32645 (German Institute for Standardization), LOD and LOQ are parameters, which describe the lowest concentration of an analyte that can be measured in an analytical process.

Molt & Telgheder (2010) have proven that it is possible to calculate the limits of detection and quantitation according to the DIN 32645 by the help of calibration measurements and the resulting regression curve, a standard procedure in instrumental analytics. Shortly, primary stock solutions were prepared in eluent A (6 mM tributylamine aqueous solution adjusted to pH 6.2 with 6 mM acetic acid) at a concentration of 1 g L^{-1} for each metabolite. From these solutions, a stock solution of 80 µM standard mixture was prepared. Aliquots of this standard mixture were used for calibration standards. A serial dilution of the standard mixture with 37 concentrations ranging from 1 nM to 10 µM was analyzed (1, 2, 3, 4, 5, 6, 7, 8, 9, 10, 20, 30, 40, 50, 60, 70, 80, 90, 100, 200, 300, 400, 500, 600, 700, 800, 900, 1000, 2000, 3000, 4000, 5000, 6000, 7000, 8000, 9000, 10000 nM). The dilution was controlled gravimetrically to calculate the exact concentration of each analyte in every sample. Calibration curves were later obtained by plotting the logarithm of the peak area against the logarithm of the concentration of each metabolite. Linear regression was used to fit the calibration curve and the linearity for each compound was derived from it. The intra-assay precision was evaluated by determination of a

QC sample at 5 µM (n=3) in one analytical batch. The precision is expressed as coefficient of variation (CV) in percent. The limit of detection and the limit of quantification (LOQ) were determined with the Excel-based program Dintest of the University Heidelberg according to the calibration method in DIN 32645 guidelines.

4.8 Validation of Metabolic Datasets by Energetic Constraints

Metabolic datasets were validated by the comparison of mass action ratio's (MAR's) of reaction and their corresponding equilibrium constant (K_{eq}) (Wittmann et al. 2005; Taymaz-Nikerel et al. 2009). The ratio (MAR/K_{eq}) of the mass action ratio to the equilibrium constant equals the distance of a reaction from its state of equilibrium, thereby allowing determination of the behavior of reaction networks (Hofmeyr & Cornish-Bowden 2000). A low MAR/K_{eq} ratio indicates a driving force into the forward direction, while the reverse reaction will be favored at values above 1.

In addition, Gibbs energies (ΔG) were calculated for every reaction by application of the MAR's to Equation 5.1.

$$\Delta G = \Delta G^{\circ\prime} + RT\ln(MAR) \qquad \text{(Equation 5.1)}$$

This equation allows not only to determinate the direction in which the particular mixture of metabolites will be able to react according to its driving force, but also shows if the driving force is strong enough to overcome the standard Gibbs energy $\Delta G^{\circ\prime}$ in the first place (Cornish-Bowden 1981). Therefore, not only the driving force but also the Gibbs energy was calculated for every reaction to validate the obtained datasets. Negative Gibbs energy implies a forward directed reaction, while positive values indicate that the reaction will proceed backwards under the investigated conditions.

In order to render the metabolic reaction network as precisely as possible, selected assumptions were made. The intracellular concentration of carbon dioxide was estimated to be 0.0133 mM in the culture broth by the utilization of Henry's law at a partial pressure of CO_2 of 0.039 kPa. Followed by the assumption of equally high concentration inside and outside of the cells this allowed computing the MAR of reactions which included the release or incorporation of a carbon dioxide molecule. Changes of the assumed carbon dioxide concentration by a 1000-fold did not affect the reactional direction, thus showing the robustness of the estimated carbon dioxide concentration. The concentration of coenzyme A was set to be 25 % of the measured AcCoA level (Jackowski & Rock 1986; Boynton et al. 1994; Chohnan et al. 1997). The concentration of FADH was set to 1 nmol/g_{CDW}, assuming that the FADH/FAD ratio is similar to the NADH/NAD ratio in viable cells.

5 Results and Discussion

5.1 Method Development and Optimization

It was the major goal of this work, to develop a fast and sensitive quantification method for metabolites from central carbon metabolism by coupling UHPLC and tandem mass spectrometry. As mentioned above, the development had to consider a number of successive steps, which had to be evaluated and optimized (Figure 5.1).

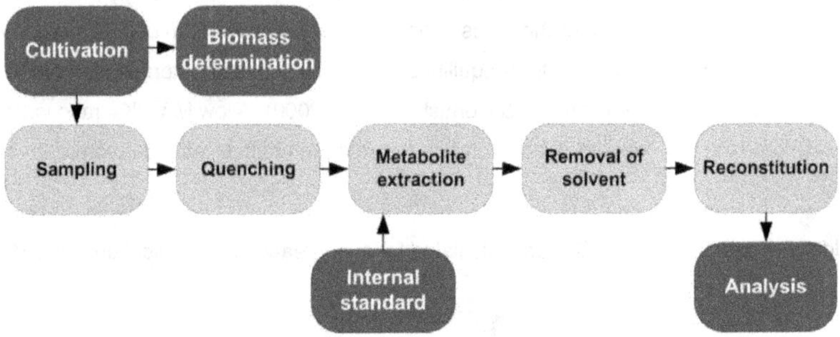

Figure 5.1: Metabolomics workflow. Figure adapted from Lämmerhofer & Weckwerth (2013)

In detail, the mass spectrometric set up had to be optimized for compounds of interest to gain maximal sensitivity. The assumed fragmentation patterns were later verified by measurement of $U^{13}C$-labeled cell extracts. Subsequently, the chromatographic separation efficiency should be optimized, in order to achieve maximal separation in a short period of time and in the presence of a biological matrix. After the analytical method had been established, the protocols of sample processing, quenching of cellular metabolism and extraction of metabolites had to be optimized, respectively.

5.1.1 Improvement of the Mass Spectrometric Set-Up for Metabolite Analysis

As specified above, multiple reaction monitoring (MRM) is a highly specific and sensitive measurement mode in mass spectrometry. In order to enable unambiguous identification of the metabolites of the central carbon metabolism important parameters had to be estimated for every compound of interest. These compound specific parameters were the *m/z*-ratios of the dominant molecular ion and of fragment ions. Afterwards, the declustering potential (DP), the collision energy (CE) and the cell exit potential (CXP) needed to be optimized to enable optimal ionization, fragmentation and detection. For this purpose, solutions of chemically pure metabolites in water (500 ng/mL) were prepared. The solutions were injected into the mass spectrometer individually via a syringe pump. A Q1 scan was used to identify the molecular ion and to determine its optimal DP. Afterwards, the fragmentation pattern of the molecular ion was measured via a product ion scan. Subsequently, the most prominent fragment was chosen. The CE, which resulted in the highest fragment ion intensity, was determined. If possible, the two most prominent transitions of every metabolite were optimized to be considered for later measurements. In the last step, the obtained parameters were applied to a MRM measurement for each metabolite in order to identify its optimal CXP. The consecutive optimization steps were conducted for different metabolites (Table 5.1) of the central carbon metabolism. These measurements yielded thoroughly fine-tuned mass spectrometer settings that enabled maximal sensitivity. One of the transitions, in most cases the main product ion, was taken for quantification (qualifier), while the other one was used as identifier. As negative ionization mode was used, the dominant molecular ion typically was $[M-H]^-$. Table 5.1 depicts the obtained mass-to-charge (*m/z*) ratios of the molecular ion $[M-H]^-$, the corresponding main product ion and the optimized potentials for each investigated metabolite. In all cases, the $[M-H]^-$ ion was found as molecular ion, except for acetyl-coenzyme A and succinyl-coenzyme A, which were ionized twice, resulting in a dominant molecular ion at a *m/z*-ratio of $[M-2H]^{2-}/2$. Main product ions of phosphorylated compounds were found at *m/z*=79 and *m/z*=97, reflecting the fragments $[PO_3]^-$ and $[H_2PO_3]^-$, respectively. Similar fragmentation patterns were observed for structural isomers, such as the hexose phosphates and pentose phosphates. These required a later chromatographic separation step prior to mass spectrometric analysis as the MS analysis alone could not be used for differentiation. For organic acids, decarboxylation and/or the loss of water were the most prominent fragmentation patterns.

Table 5.1: Transitions of 33 unlabeled and U^{13}C-labeled metabolites with the corresponding declustering potential (DP), collision energy (CE) and cell exit potential (CXP) used for multiple reaction monitoring (MRM).

Compound	Molecular Ion		Fragment Ion		Structure of Cleavage	DP	CE	CXP
	Std.	[U^{13}C]	Std.	[U^{13}C]				
	[*m/z*]	[*m/z*]	[*m/z*]	[*m/z*]		[V]	[V]	[V]
GA	72.8	74.8	44.9	45.9	$[COOH]^-$	-50	-14	-13
PYR	86.8	89.8	43.0	45.0	$[M-CO_2]^-$	-38	-14	-19
FUM	114.8	118.8	71.0	74.0	$[M-CO_2]^-$	-35	-10	-12
SUC	116.9	120.9	99.1	103.1	$[M-H_2O]^-$	-36	-20	-13
MAL	132.9	136.9	115.0	119.0	$[M-H_2O]^-$	-41	-15	-10
AKG	144.9	149.9	101.0	105.0	$[M-CO_2]^-$	-45	-12	-14
PEP	166.8	169.8	78.9	78.9	$[PO_3]^-$	-31	-22	-12
DHAP	169.0	172.0	97.0	97.0	$[H_2PO_4]^-$	-54	-12	-7
GAP	169.0	172.0	97.0	97.0	$[H_2PO_4]^-$	-35	-16	-9
2PG	184.9	187.9	79.0	79.0	$[PO_3]^-$	-24	-58	-13
3PG	184.9	187.9	79.0	79.0	$[PO_3]^-$	-67	-64	-8
ISOCIT/CIT	190.8	196.8	129.0	134.0	$[M- H_2O, -CO_2]^-$	-60	-20	-18
ISOCIT	190.9	196.9	72.9	74.9	$[C_2HO_3]^-$	-50	-32	-27
R5P	228.8	233.8	79.0	79.0	$[PO_3]^-$	-55	-54	-10
X5P	229.1	234.1	78.9	78.9	$[PO_3]^-$	-45	-50	-8
RIBU5P	229.1	234.1	97.0	97.0	$[H_2PO_4]^-$	-44	-16	-16

Compound	Molecular Ion		Fragment Ion		Structure of Cleavage	DP	CE	CXP
	Std.	[$U^{13}C$]	Std.	[$U^{13}C$]				
	[*m/z*]	[*m/z*]	[*m/z*]	[*m/z*]		[V]	[V]	[V]
KDPG	257.2	263.2	97.1	97.1	$[H_2PO_4]^-$	-25	-26	-7
G6P	259.1	265.1	97.0	97.0	$[H_2PO_4]^-$	-58	-23	-17
F6P	259.1	265.1	97.0	97.0	$[H_2PO_4]^-$	-86	-17	-8
F1P	259.3	265.3	97.0	97.0	$[H_2PO_4]^-$	-32	-22	-6
6PG	275.0	281.0	79.0	79.0	$[PO_3]^-$	-73	-68	-7
S7P	289.1	296.1	96.9	96.9	$[H_2PO_4]^-$	-10	-30	-13
FBP	339.2	345.2	97.0	97.0	$[H_2PO_4]^-$	-80	-36	-14
AMP	346.0	356.0	79.0	79.0	$[PO_3]^-$	-70	-56	-6
AcCoA	403.6	415.3	78.8	78.8	$[PO_3]^-$	-45	-94	-11
ADP	426.0	436.0	79.0	79.0	$[PO_3]^-$	-76	-82	-11
SucCoA	432.8	445.3	79.1	79.1	$[PO_3]^-$	-11	-116	-10
ATP	506.2	516.2	78.9	78.9	$[PO_3]^-$	-99	-130	-19
NAD	662.2	683.2	540.1	555.1	[M-Nicotinamide]$^-$	-92	-24	-34
NADH	664.2	685.2	78.9	78.9	$[PO_3]^-$	-120	-136	-9
NADP	742.0	763.0	620.1	635.1	[M-Nicotinamide]$^-$	-75	-30	-29
NADPH	744.0	765.0	79.0	79.0	$[PO_3]^-$	-132	-116	-5
FAD	784.0	811.0	437.2	454.2	$[C_{17}H_{18}N_4O_8P]^-$	-150	-38	-25

As desired, all 33 metabolites of interest could be successfully tuned in the negative ionization mode. The assumed structures of cleavage were later confirmed by analysis of [$U^{13}C$] labeled metabolite extracts of *C. glutamicum* (see chapter 0). Here, transitions, which were corrected for the number of labeled carbon atoms, resulted in peaks at retention times similar to the measurement of the unlabeled metabolite mix during tuning. It became obvious that the optimization of DP, CE and CXP resulted in signal intensities, which were more than 10-fold increased, for selected transitions. Thus, the successful tuning process enabled the detection of even smallest abundancies of the metabolites in biological samples.

5.1.2 Chromatographic Separation of Standard Mixtures

During the initial tuning studies, it turned out that certain metabolites mimicked the fragmentation patterns of other metabolites, even though both compounds possessed different molecular masses and thus different molecular ions. This behavior obviously reflected collision induced dissociation (CID), which takes place in the ion source or in the interface of the LC-MS. A prominent example was the observed molecule ion of ATP, exhibiting a *m/z*-ratio of 506 and a transition of *m/z*=506→79 under optimal conditions. If, however, ATP becomes dephosphorylated in the ion source by CID before entering the first quadrupole, this will result in an additional molecular ion with *m/z*=426, which is then decomposed in the collision cell to the main product ion *m/z*=79 ($[PO_3]^-$). The result is the measured transition (426→79), thereby mimicking the transition of ADP, even though no ADP is present in the sample. This can be seen in the MRM chromatogram (Figure 5.2A). Here the mentioned in-source dephosphorylation of

ATP resulted in a second peak of the ADP specific transition (426→79) at the retention time of ATP (t=14.2 min). Additionally, CID was observed for MAL and FUM as well as SUC and αKG (Figure 5.2B). Likewise, the sugar-phosphate FBP caused a false peak in the extracted ion chromatogram (XIC) of the transition of dihydroxy-acetone-phosphate and glyceraldehyde-3-phosphate. Vice versa, the transition *m/z*=339→97 (FBP) showed two additional peaks at the retention times of DHAP and GAP, respectively. This finding could be caused by a temperature induced aldol-formation of FBP from DHAP and GAP during the ionization process. Even though it might be possible to diminish interfering signals by tuning of the ion source or interface settings (i.e. temperature, solvent, flow rate), it is unlikely that all of the undesired reactions could be suppressed to full extent, while maintaining high sensitivity. These false positive peaks show the importance of employing a powerful chromatographic method, by which metabolites like ADP and ATP can be separated. Otherwise the quantification of ADP could easily become corrupted by the CID transition of ATP.

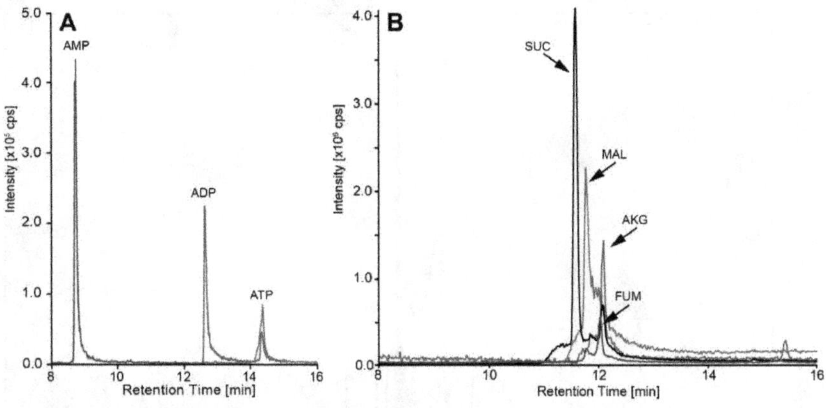

Figure 5.2: Extracted Ion Chromatogram (XIC) of metabolite measurements using the Multiple Reaction Monitoring (MRM) mode of the LC-MS/MS system. Combination of chromatography and mass spectrometry enables specific identification of AMP (346→79), ADP (426→79) and ATP (506→79) as well as SUC (117→99), MAL (133→115), AKG (145→101) and FUM (115→71) by retention time and fragmentation pattern. (A) Two peaks rise at the retention time of ATP, the metabolite specific transition of ATP (506→79) and the ADP (426→79) specific signal as a consequence of CID of ATP (506→426) in the LC-MS interface. (B) CID was also monitored for MAL, resulting in an additional FUM signal and for AKG, resulting in an additional SUC peak.

For chromatographic separation of the metabolites of interest, a reversed phase UHPLC in combination with the ion pairing reagent TBA was chosen. For the following optimization, 2 µL of a standard mixture, which contained 1 µM of each metabolite, was repeatedly injected into the LC-MS/MS system. Measurements were repeated three times for each experimental set up. In the course of method optimization, two parameters were identified, which heavily influenced separation and sensitivity of the quantification method. The first parameter, which

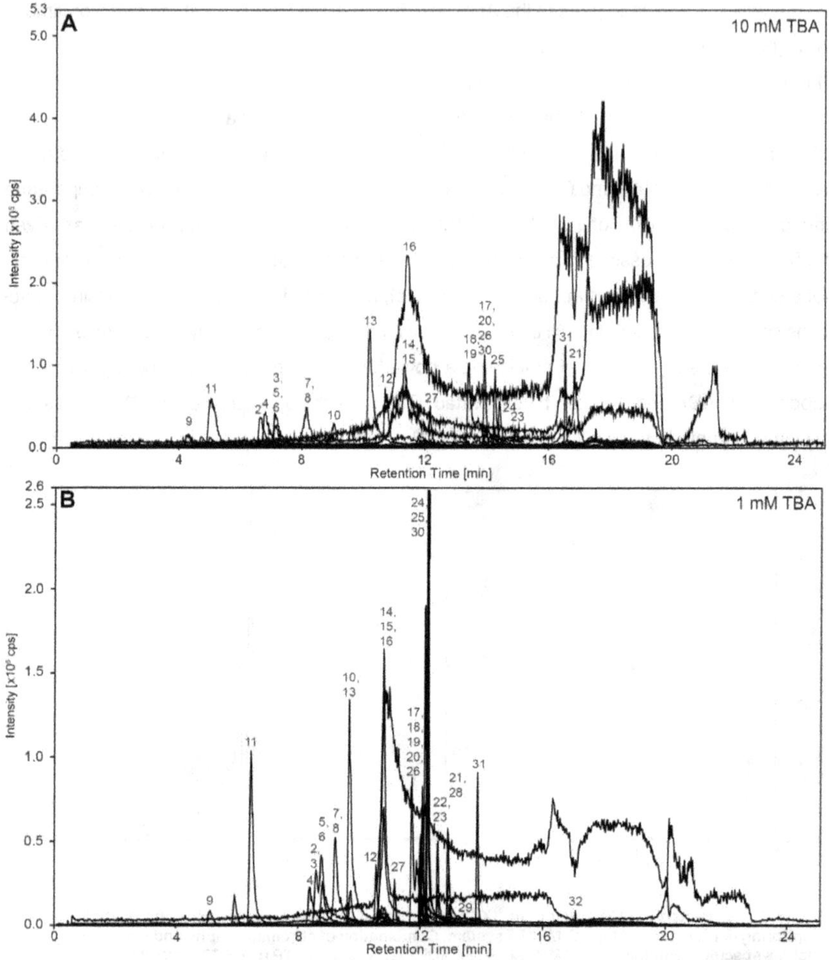

Figure 5.3: Combined selective ion chromatograms of 32 metabolites in a 1 µM standard mixture (injection volume 2 µL) at a column temperature of 25 °C. Separation was achieved by application of VisionHT UHPLC-column (Grace), while detection was performed with the QTRAP5500 LC-MS/MS (AB/Sciex). Eluents contained 10 mM TBA (A) and 1 mM TBA (B). Further details on the used LC-MS/MS method, eluent composition and gradient profile are described in the experimental part. Peak identification: (1) GA; (2) G6P; (3) GAP; (4) R5P; (5) S7P; (6) F6P; (7) RIBU5P; (8) X5P; (9) PYR; (10) DHAP; (11) NAD; (12) SUC; (13) AMP; (14) MAL; (15) FUM; (16) AKG; (17) 6PG; (18) 2PG; (19) 3PG; (20) KDPG; (21) FBP; (22) CIT; (23) ISOCIT; (24) PEP; (25) NADP; (26) ADP; (27) NADH; (28) ATP; (29) NADPH; (30) FAD; (31) AcCoA; (32) SucCoA

needed optimization, was the concentration of the ion pairing reagent. As previously mentioned, (see chapter 3.2.1), the use of ion pairing reagents is helpful to prolong the retention of metabolites by RP-columns. However, the ion pairing reagent obviously caused ion suppression and loss in sensitivity. Therefore, the effect of TBA concentrations, ranging from 1 mM to 10 mM, on separation efficiency and sensitivity were tested. The resulting chromatograms for measurements with (A) 10 mM TBA and (B) 1 mM TBA, respectively, are shown in Figure 5.3.

Obviously, the lowered TBA concentration of 1 mM led to a signal intensity, which was up to five times higher than in the presence of 10 mM TBA. Simultaneously, peaks became narrower and background noise was drastically reduced. These findings were accompanied by a still sufficient separation of most stereochemically isomeric compounds, which allowed successful differentiation by retention time. Consequently, all further experiments were conducted using 1 mM TBA. However, not all substances were fully resolved. In order to widen the elution window, while keeping the low concentrations of TBA, the column temperature was raised in steps of 5 °C. As an example, resulting chromatograms at column temperatures of 25 °C and 50 °C are shown in Figure 5.4.

Higher temperatures resulted in a broadened range of retention times for all metabolites of interest and even enhanced the separation of metabolites eluting early. Figure 5.4 A shows, that metabolites mainly elute in two large clusters at 25 °C. The first cluster at a retention time window of 8 to 10 minutes, consists of the sugar phosphates and the organic acids MAL, FUM, SUC and AKG. The second cluster eluting between 12 to 16 minutes consisted of energy metabolites, redoxequivalents and CoA esters. In Figure 5.4 B the positive effect of the higher temperature on separation power can be seen, as the peaks elute well separated. Higher temperatures result in a decrease in viscosity of the mobile phase and an increased analyte diffusivity (Greibrokk & Andersen 2003). Both effects lead to a better interaction of column material, ion-pairing reagent and metabolites. Obviously, the seperation of metabolites, which were susceptible to the retaining effect of TBA, were seperated more efficiently at higher temperature. Measurements at 30 °C, 35 °C, 40 °C and 45 °C showed, that every temperature increase resulted in a better separation of the sugar phosphate cluster. Overall, the temperature increase from 25 °C to 50 °C resulted in a 2.5 times wider elution window for these compounds, thereby enhancing separation efficiency of the method. However, a further increase of the temperature was not considered, as some metabolites are known to be thermolabile. Consequently, a higher temperature might result in reduced sensitivity, a fact that was observed for PYR (Figure 5.4). Hence, all further measurements were performed at a column temperature of 50 °C.

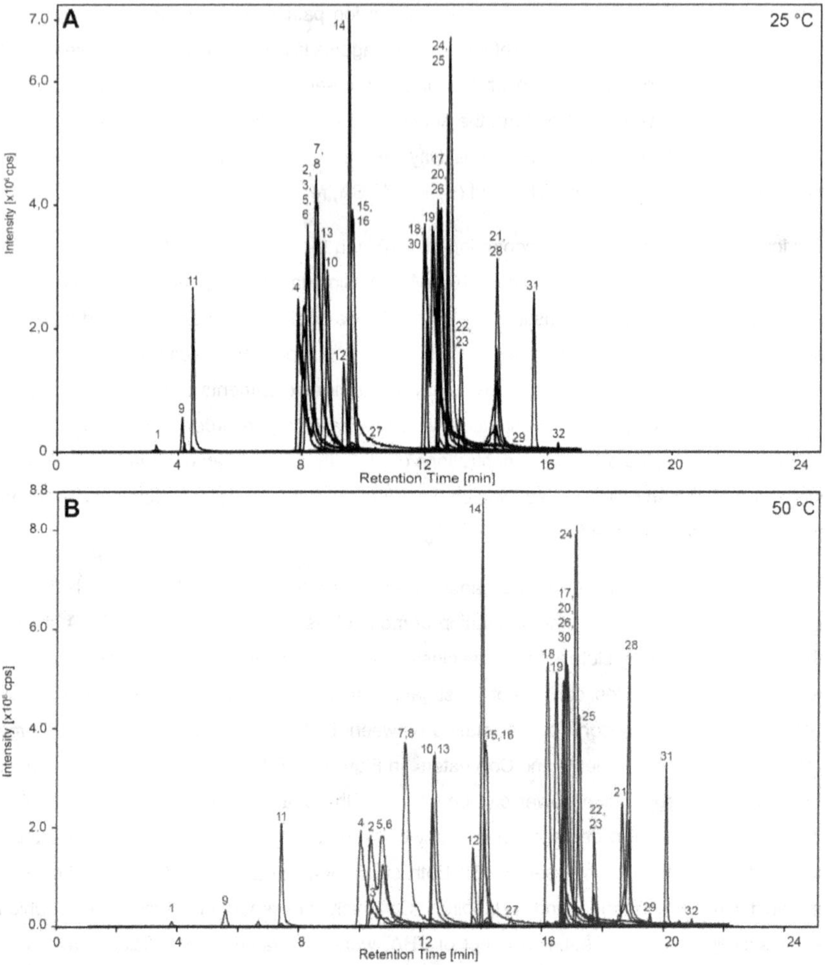

Figure 5.4: Combined selective ion chromatograms of 32 metabolites in a 1 µM standard mixture (injection volume 2 µL) using eluents containing 1 mM TBA at a column temperature of 25 °C (A) and 50 °C (B). Separation was achieved by application of VisionHT UHPLC-column (Grace), while detection was performed with the QTRAP5500 LC-MS/MS (AB/Sciex). Further details on the used LC-MS/MS method, eluent composition and gradient profile are described in the experimental part. Peak identification: (1) GA; (2) G6P; (3) GAP; (4) R5P; (5) S7P; (6) F6P; (7) RIBU5P; (8) X5P; (9) PYR; (10) DHAP; (11) NAD; (12) SUC; (13) AMP; (14) MAL; (15) FUM; (16) AKG; (17) 6PG; (18) 2PG; (19) 3PG; (20) KDPG; (21) FBP; (22) CIT; (23) ISOCIT; (24) PEP; (25) NADP; (26) ADP; (27) NADH; (28) ATP; (29) NADPH; (30) FAD; (31) AcCoA; (32) SucCoA

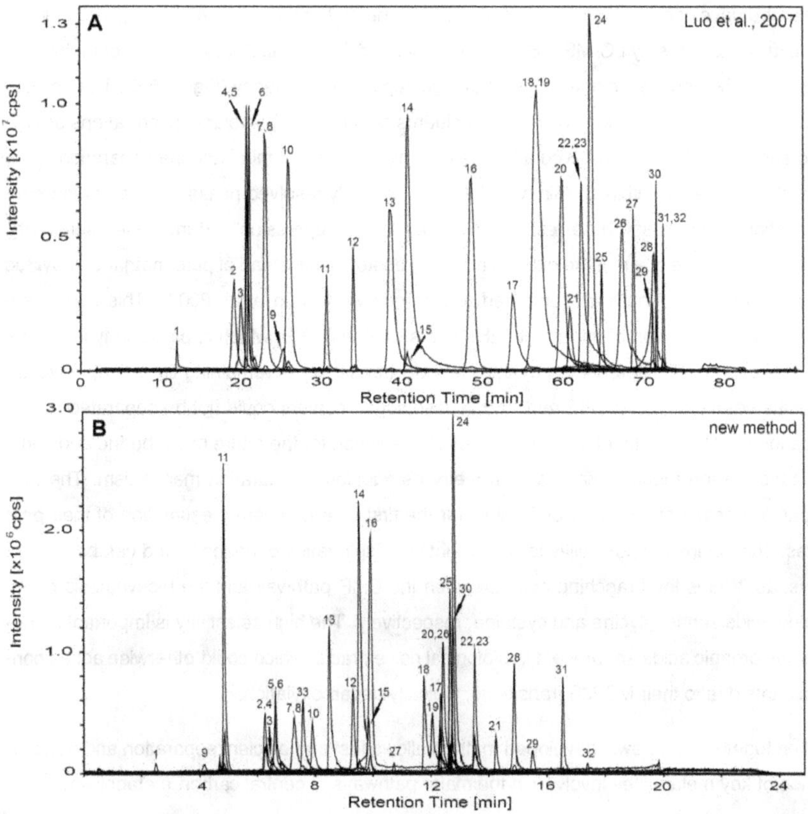

Figure 5.5: Combined selective ion chromatograms of 33 metabolites in a 1 µM standard mixture. According to the method published by Luo et al. (2007) (A) and according to the newly developed method (B). For (A) 10 µL standard mixture were injected, while for (B) 2 µL standard mixture were injected, thus resulting in substantially lower signal intensities for the novel method. For (B) separation was achieved by application of VisionHT UHPLC-column (Grace), while detection was performed with the QTRAP5500 LC-MS/MS (AB/Sciex). Further details on the used LC-MS/MS method, eluent composition and gradient profile are described in the experimental part. Peak identification: (1) GA; (2) G6P; (3) GAP; (4) R5P; (5) S7P; (6) F6P; (7) RIBU5P; (8) X5P; (9) PYR; (10) DHAP; (11) NAD; (12) SUC; (13) AMP; (14) MAL; (15) FUM; (16) AKG; (17) 6PG; (18) 2PG; (19) 3PG; (20) KDPG; (21) FBP; (22) CIT; (23) ISOCIT; (24) PEP; (25) NADP; (26) ADP; (27) NADH; (28) ATP; (29) NADPH; (30) FAD; (31) AcCoA; (32) SucCoA; (33) F1P

Next, the developed approach was compared to the state of the art, i.e. the analysis of intracellular metabolites by LC-MS/MS recently described (Luo et al. 2007). The novel method allowed a 70% reduction in analysis time at high separation efficiency (Figure 5.5). The application of a UHPLC separation column, with eluents containing equimolar concentrations of TBA and acetic acid (pH 6.2), allowed a flow rate as high as 400 µL min^{-1} and the separation of the analytes of interest within 28 min with defined and highly resolved peaks. The rather low concentration of TBA used here resulted in reduced ion suppression and increased sensitivity. Moreover, the use of a mixture of 50 % aqueous acetonitrile instead of pure methanol provided stronger elution strength as compared to previous work (Luo et al. 2007). This further enhanced the separation of the structural isomers G6P and F6P, which is particularly important as they cannot be differentiated by their MS/MS transitions. A remaining limitation, however, was the co-elution of RIBU5P and X5P. These sugar isomers could not be separated by application of TBA (Balcke et al. 2011). A possible solution for the future might be the additional utilization of ion mobility mass spectrometry as additional separation mechanism. The fully separated peaks of 2PG and 3PG allow for the first time a separate estimation of their pool sizes. This opens the possibility to get insights into split ratios of anabolic and catabolic reactions, as 3PG is the branching point between the EMP pathway and the biosynthesis of the amino acids serine, glycine and cysteine, respectively. The high selectivity is important as numerous organic acids are present in biological cell extracts, which could otherwise act as contaminants due to their MS/MS transitions, typically decarboxylation.

Taken together, the newly developed method allows fast and efficient separation and quantification of key metabolites involved in the major pathways of central carbon metabolism.

5.1.3 Chromatographic Separation of Cellular Extracts

After optimization of the chromatographic separation for standard mixtures, an important next step was to evaluate the influence of the biological matrix of real samples on the separation efficiency. For this purpose, a standard mixture was spiked with ubiquitously ^{13}C-labeled extract of *C. glutamicum* (see chapter 4.6.3). For evaluation 2 µL of the obtained mixture was injected into the LC-MS/MS system. In short, the spiking with labeled extract allowed differentiating between the analyte from the biological sample and the standard mixture, thus enabling a fast identification of the metabolites and their corresponding peaks in the presence of the biological matrix.

First measurements with 1 mM TBA and 1 mM AA in the eluent revealed a drastic effect of biological matrix on separation efficiency (Figure 5.6 A). Obviously, a concentration of 1 mM TBA was optimal to separate pure standard mixtures, but appeared insufficient to achieve adequate separations for cell extracts. The concentration of TBA seemed not high enough to interact with the much higher number of molecules, contained in the biological samples. Consequently, the metabolites of interest were not separated as effectively as in the pure standard mixture (Figure 5.5 B), thereby resulting in a rather clustered elution between 11 to 15 min of analysis. In addition, the signal to noise ratio decreased. Hence, metabolites with lower signal intensities, such as GA and PYR, were strongly overlaid by background noise. To compensate this undesired behavior the concentration of TBA was increased stepwise. This finally resulted in optimal separation at around 6 mM TBA in the mobile phase (Figure 5.6 B). Comparable to the separation of standard mixture with 1 mM TBA (Figure 5.5 B), the analytes eluted regularly from the column. Differentiation of stereo isomeric substances, like G6P and F6P, could be regained. Additional measurements with 10 mM of TBA, however, caused again reduced signal intensity. All following measurements were therefore performed using 6 mM TBA and AA to allow optimal distinction between single metabolites, at minimal loss of sensitivity due to ion suppression.

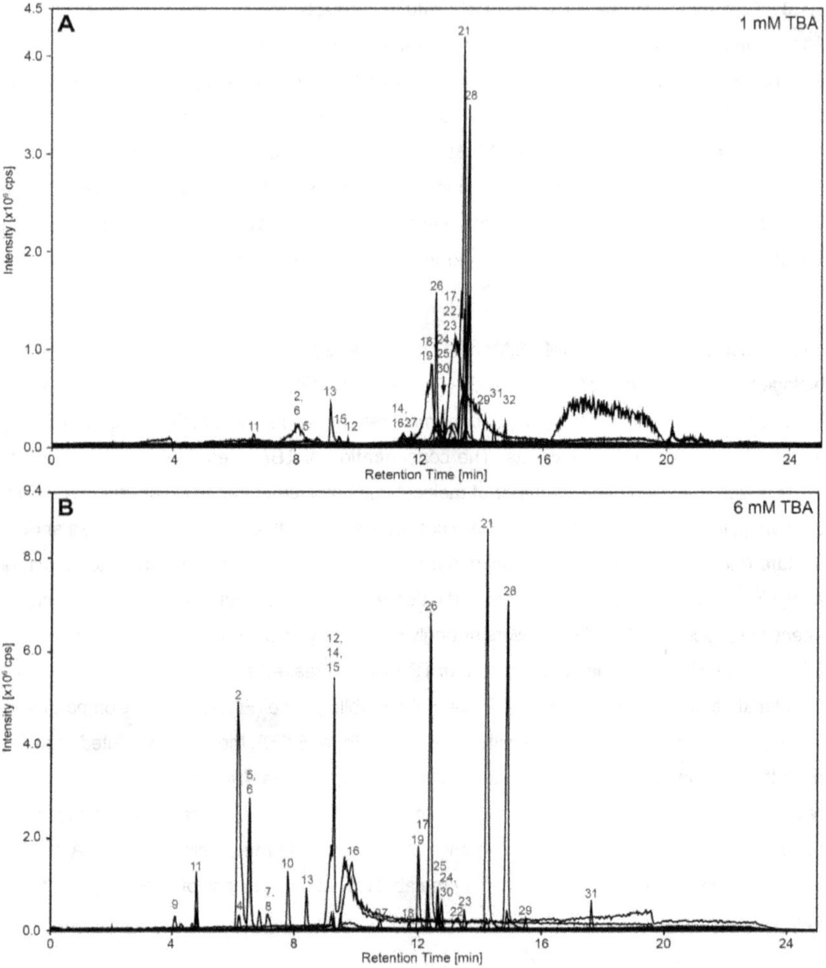

Figure 5.6: Combined selective ion chromatograms of 32 metabolites in a 1 µM standard mixture (injection volume 2µL) spiked with labeled cell extract (1:1) at a column temperature of 50 °C. Using eluents containing 1 mM TBA (A) and using eluents containing 6 mM TBA (B). Separation was achieved by application of VisionHT UHPLC-column (Grace), while detection was performed with the QTRAP5500 LC-MS/MS (AB/Sciex). Further details on the used LC-MS/MS method, eluent composition and gradient profile are described in the experimental part. Peak identification: (1) GA; (2) G6P; (3) GAP; (4) R5P; (5) S7P; (6) F6P; (7) RIBU5P; (8) X5P; (9) PYR; (10) DHAP; (11) NAD; (12) SUC; (13) AMP; (14) MAL; (15) FUM; (16) AKG; (17) 6PG; (18) 2PG; (19) 3PG; (20) KDPG; (21) FBP; (22) CIT; (23) ISOCIT; (24) PEP; (25) NADP; (26) ADP; (27) NADH; (28) ATP; (29) NADPH; (30) FAD; (31) AcCoA; (32) SucCoA

Additionally, the effects caused by of the biological matrix illustrated the importance of using multiple transitions for unambiguous identification of a metabolite. As can be seen, biological samples contained metabolites with similar molecular ions and fragmentation patterns which cause unspecific peaks, so called matrix effects (Figure 5.7).

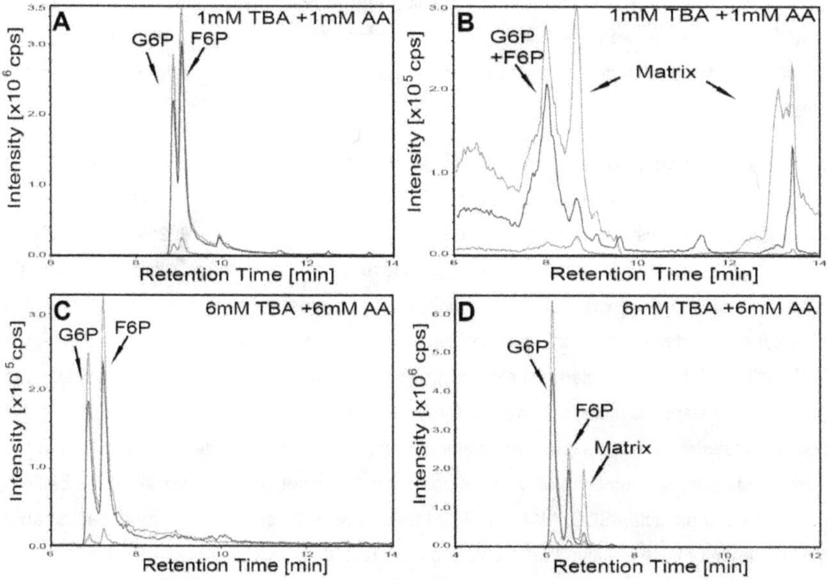

Figure 5.7: Extracted Ion Chromatogram (XIC) of G6P, F6P and F1P in a standard mixture (A and C) and cellular extracts of *C. glutamicum* ATCC 13032 (B and D). The colored lines represent the different transitions used for the measurement of G6P and F6P: Green: 259.1→78.9, blue: 259.1→97, red: 259.1→138.8. LC-MS/MS conditions: Eluents were prepared equimolar: (A) and (B) 1mM TBA mixed with 1mM AA, pH 6.2; (C) and (D) 6mM TBA mixed with 6mM AA, pH 6.2. All other LC-MS/MS conditions were left unchanged.

Figure 5.7 depicts the three extracted ion chromatograms (XIC's) of G6P/F6P specific transitions in a standard mixture of G6P and F6P and a cellular extract, whereby the mobile phase contained 1 mM and 6 mM TBA, respectively. Figure 5.7 A and C depict measurements of the standard mixture and Figure 5.7 B and D show chromatograms of cellular extracts. For the standard mixture, a concentration of 1 mM TBA was sufficient to fully separate G6P and F6P (Figure 5.7A). Higher TBA concentration led to even better separation of the isomers, but somewhat reduced the overall intensity due to peak broadening and ion suppression (Figure 5.7C). The analysis of a biological sample, however, strongly differed. The measurement at 1 mM TBA showed two broad peaks at the correct retention times for G6P and F6P with low signal intensity (Figure 5.7B), while 6 mM TBA resulted in three defined and sharp peaks at high signal intensity (Figure 5.7D). A closer look to the ratio between the transitions for the cell extract revealed that the ratio for the peak eluting at 9 min in Figure 5.7 B and peak eluting at 7 min in Figure 5.7 D did not resemble the expected ratio for G6P and F6P. These peaks were obviously matrix derived signals. The results emphasized the importance of multiple transitions for the clear identification of chromatographic peaks by mass spectrometry. It was therefore decided to always use one quantifying transition and at least one identifying transition. The quantifying transition was used to quantify the metabolite concentration, while the ratio between quantifying and identifying transition was used to identify potentially falsifying matrix effects.

5.1.4 Verification of Ion Fragmentation Patterns

Isotope dilution mass spectrometry has proven to greatly enhance the robustness of quantitative analysis of atoms or molecules via mass spectrometry (Heumann 1992; Wu et al. 2005). Thus, *C. glutamicum* cells, grown on [$^{13}C_6$]-glucose as sole carbon source, were used to obtain a fully ^{13}C-labeled cell extract with the metabolites of interest (see chapter 4.6.3). Figure 5.8 shows the extracted ion chromatograms of all 33 metabolites in the standard mixture compared to the XIC's of the ^{13}C labeled cell extract. Overall, the retention time for all metabolites was stable for the standard mixture as well as for the cell extract. Thus, the intracellular matrix and possible residues of the medium had no effect on the retention time. The high specificity of the MRM mode enabled detecting only single peaks for the majority of metabolites in the cell extract. The organic acids SUC, MAL and FUM as well as ADP showed additional peaks, as a consequence of false positive transitions and crosstalk, as discussed above (Figure 5.2). The chromatogram for NADH revealed a second peak at the retention time for NADP (12.66 min), i.e. a false positive signal resulting from dephosphorylation of NADP. Furthermore, NADPH exhibited an additional peak at the retention time of NADP (12.66 min). This observation could be explained by the mass spectrometric resolution. All measurements were conducted in unit resolution, which allowed a window of one Da below and above of the actual atomic mass, which could lead to false positive signals for substances with almost equal molar masses. The isomers RIBU5P and X5P could not be separated, neither in the extract nor in the standard mixture. It was nevertheless possible to distinguish between the isomeric compounds ISOCIT and CIT, by retention time and their specific fragmentation pattern. In contrast to all other isomers applied in this method, ISOCIT and CIT could be identified by their specific transitions of m/z=191→73 and m/z=191→129, respectively. The differentiation between these two isomers is interesting for metabolic research, as it provides access on reactions from citrate into the biosynthesis of fatty acids (Ratledge 2004).

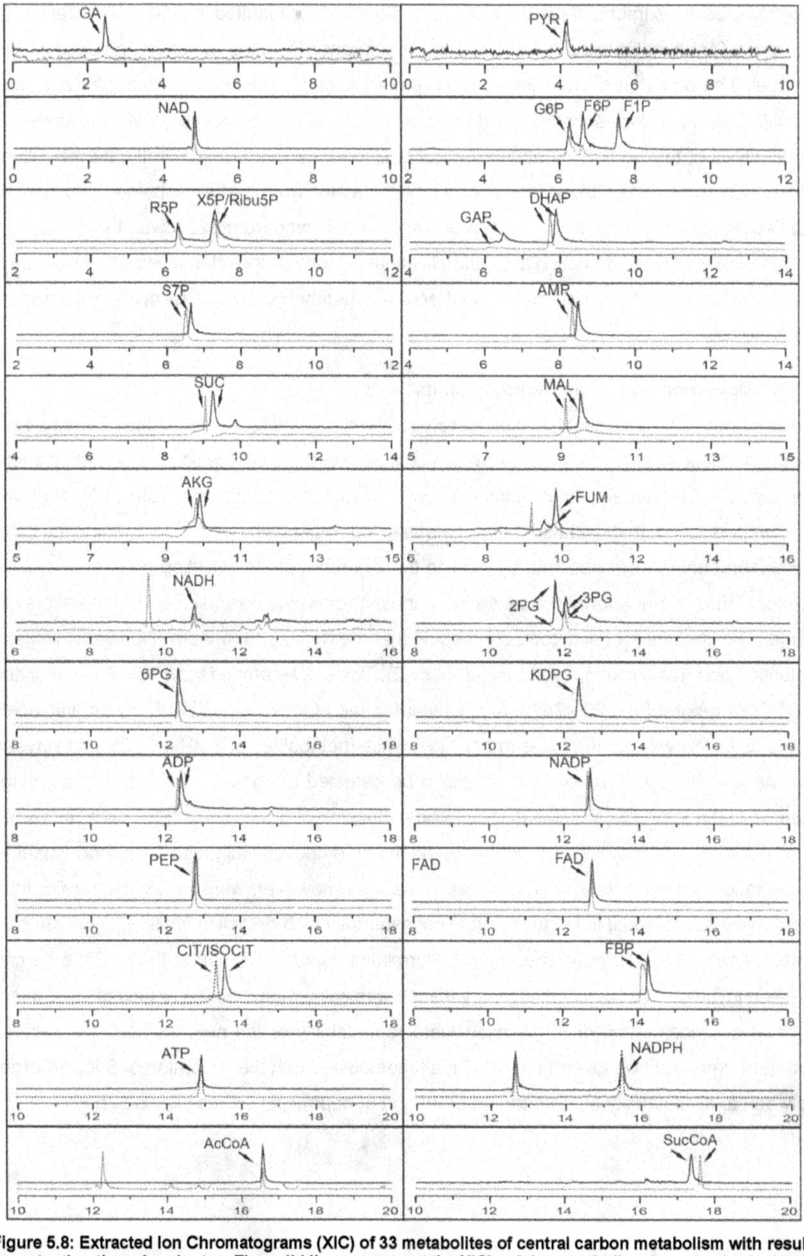

Figure 5.8: Extracted Ion Chromatograms (XIC) of 33 metabolites of central carbon metabolism with resulting retention times in minutes. The solid lines represent the XIC's of the metabolites in a standard mixture and the dashed lines stand for the XIC's of the corresponding metabolites in extracts of the gram-positive soil bacterium *Corynebacterium glutamicum* ATCC 13032.

The ^{13}C-labeled extract and the unlabeled standard mixture resulted in peaks at similar retention times for all metabolites of interest despite different *m/z*-ratios, which are a consequence of labeled carbon atoms. The fragmentation patterns, which had been deducted for each metabolite from their corresponding product ions during the tuning process, were confirmed by the analysis of the labeled metabolite extract of *C. glutamicum*. Consequently, the transitions can now be modified to allow detection of labeling states from partially labeled to completely labeled metabolites of interest. The ^{13}C-labeled internal standard now allowed the precise determination of limits of detection and quantification. Moreover, the internal standard was afterwards applied for exact quantification of intracellular metabolites by isotope dilution mass spectrometry.

5.1.5 Detection and Quantification Limits

Limits of detection and quantification were evaluated according to DIN 32645. To obtain the required calibration curves, a serial dilution of a standard mixture, which contained synthetic standards of the 33 metabolites of interest at 37 different concentrations, from 1 nM to 10 µM, was prepared and analyzed using the ^{13}C-labeled internal standard. From experimental data, calibration curves were obtained by plotting the logarithm of the resulting peak area against the logarithm of the known concentration. Linear regression was used to fit the calibration curve. The linearity for each compound could then be derived from there. Generally, limits of detection and quantification were low (Table 5.2). The LOD's ranged from 10 - 2210 fmol and the LOQ's ranged from 20 - 7470 fmol. Obviously, the LOD's and LOQ's of the organic acids and of SucCoA were found to be rather high, while metabolites of EMP and PP pathway, as well as energy and redox equivalents could be detected in traces. This finding is a consequence of the available transitions. The large molecules of glycolysis and PP pathway exhibited highly specific transitions, which resulted in low detection limits due to a good signal to noise ratio. The transitions of the smaller molecules, however, were rather unspecific, thus resulting in a bad signal to noise ratio and consequently high detection limits. Most of the calibration graphs showed good linearity with correlation factors (R^2) higher than 0.9796 as can be seen in Table 5.2. The coefficient of variation was calculated from the results of three injections at a concentration of 5 µM to evaluate repeatability of the new method. The relative standard deviation was lower than 5 % for all metabolites with the exception of SucCoA probably due to its instability at room temperature and at neutral pH values (Table 5.2).

Table 5.2: Detection and quantification limits of 33 metabolites of the central carbon metabolism. Retention Time (RT), range of regression, relative standard deviation (R^2) of the 5 µM standard mixture (n=3) and Limits of Detection (LOD) and Quantification (LOQ) as determined by the established LC-MS/MS method (10 µL injection volume).

Compound	RT	Regression		Repeatability	LOD	LOQ
	[min]	Dynamic Range	R^2	[n=3, CV %]	[fmol]	[fmol]
GA	2.47	2000 fmol - 1000 nmol	0.9973	1.77	2210	7470
PYR	4.15	500 fmol - 100 nmol	0.9999	0.60	810	2870
FUM	9.84	400 fmol - 100 nmol	0.9935	0.61	340	1120
SUC	9.31	400 fmol - 100 nmol	0.9938	1.02	700	2110
MAL	9.52	200 fmol - 100 nmol	0.9999	4.06	130	460
AKG	9.91	50 fmol - 100 nmol	0.9999	0.95	60	230
PEP	12.80	10 fmol - 100 nmol	0.9998	1.96	10	20
DHAP	6.50	10 fmol - 100 nmol	0.9968	2.22	20	80
GAP	7.91	50 fmol - 100 nmol	0.9998	0.32	50	160
2PG	11.76	10 fmol - 100 nmol	0.9999	3.56	10	20
3PG	12.04	100 fmol - 100 nmol	0.9995	3.83	80	280
CIT	13.50	500 fmol - 100 nmol	0.9939	2.17	660	1950
ISOCIT	13.51	10 fmol - 100 nmol	0.9975	2.33	30	110
R5P	6.29	50 fmol - 100 nmol	0.9996	2.35	70	240
X5P	7.28	10 fmol - 100 nmol	0.9987	3.12	20	50
RIBU5P	7.28	10 fmol - 100 nmol	0.9997	2.02	10	20
KDPG	12.30	50 fmol - 100 nmol	0.9973	3.77	40	120
G6P	6.26	10 fmol - 100 nmol	0.9996	0.70	10	30
F6P	6.61	10 fmol - 100 nmol	0.9991	3.13	20	70
F1P	7.57	30 fmol - 100 nmol	0.9999	0.38	10	20
6PG	12.34	10 fmol - 100 nmol	0.9986	2.38	20	60
S7P	6.63	10 fmol - 100 nmol	0.9994	1.56	30	90
FBP	14.23	10 fmol - 100 nmol	0.9999	2.55	30	110
AMP	8.49	40 fmol - 100 nmol	0.9996	2.29	10	30
AcCoA	16.54	10 fmol - 100 nmol	0.9991	2.80	10	50
ADP	12.42	10 fmol - 100 nmol	0.9999	1.56	10	20
SucCoA	17.41	3000 fmol - 100 nmol	0.9852	15.02	1460	3890
ATP	14.87	10 fmol - 100 nmol	0.9994	5.09	10	50
NAD	4.85	10 fmol - 100 nmol	0.9997	3.01	10	20
NADH	10.71	10 fmol - 100 nmol	0.9990	7.64	10	20
NADP	12.66	10 fmol - 100 nmol	0.9997	0.99	10	30
NADPH	15.47	50 fmol - 100 nmol	0.9998	2.21	50	160
FAD	12.77	10 fmol - 100 nmol	0.9963	2.71	30	90

PYR and GA were overlaid by high background noise, as a consequence of their small and unspecific fragment ions. The product ions of PYR and GA, *m/z*=43 and *m/z*=44.9, additionally exhibited low intensities. The combination of both effects resulted in relatively high LODs and LOQs. Taking all metabolites into account, most LODs and LOQs were similar to previous estimates using LC-MS/MS by Luo et al. (2007) and up to 100-fold lower than reported for previous LC-MS/MS methods (van Dam 2002; Huck et al. 2003). Most LODs were below 100 fmol. This enables the detection of even traces of the metabolites in biological samples.

Consequently, cultures could be sampled at low cell concentrations without any loss of information. Moreover, the high dynamic range of the quantitation method now enabled direct measurement of cell extracts without further need for concentration or dilution, thus eliminating possible error sources during sample pretreatment.

5.1.6 Optimization of Sample Pretreatment

As mentioned before, sample pretreatment is a crucial step of metabolite analysis (see chapter 3.3) and appropriate protocols are the key to high reproducibility and correctness of data. It was therefore important to verify that quenching during sampling was fast enough to prevent falsification of metabolite data. This was studied by a simple experiment. In short, 1 g/L [$^{13}C_6$] glucose was added to the quenching solution prior to sampling. The labeling analysis of intracellular metabolites then allowed following ongoing glucose uptake during quenching. Figure 5.9 depicts the relative ^{13}C-labeling of intracellular G6P and FBP. The experiment was conducted with *E. coli* K12 DSM 2670 cultures, which were grown on glucose. For the first experiment, *E. coli* was grown in batch mode, while the second culture was grown under growth limiting fed-batch conditions. While the cells were growing with μ_{max} during the batch experiment, the fed-batch process allowed only a reduced growth rate of μ=0.17 h^{-1}. In both cases, significant labeling enrichment was observed for intracellular metabolites.

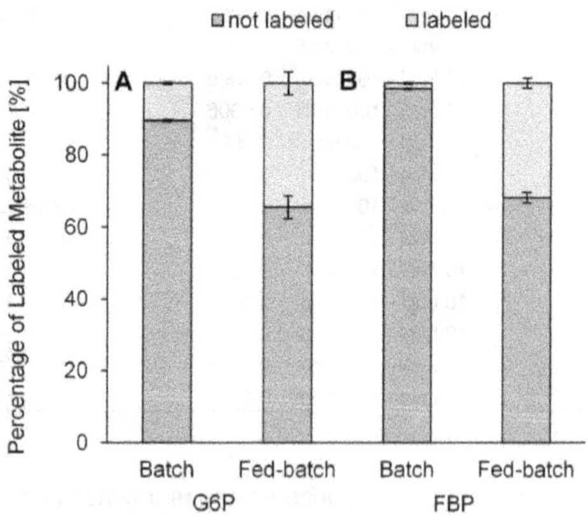

Figure 5.9: Quenching efficiency of pre-cooled aqueous methanol (60% v/v) under batch and fed-batch conditions. *E. coli* K12 DSM 2670 was grown in a bioreactor using defined medium with unlabeled glucose as the sole carbon source. The quenching solution, which was used to stop cell metabolism during the sampling process, was spiked with 1 g/L [$^{13}C_6$]-glucose. Analysis of the resulting intracellular labeling state of (A) G6P and (B) FBP enabled evaluation of the glucose uptake and thus the quenching efficiency. Analysis was performed in duplicate; mean values from there are shown.

The relative fraction of ^{13}C-labeled G6P after quenching of a fed-batch derived sample was almost three times higher than that from a batch cultivation sample (Figure 5.9), a trend also observed for FBP. This observation can be explained by the relative fraction of the labeled glucose in the quenching solution in comparison to the residual unlabeled glucose remaining in the surrounding medium. The remaining glucose concentration of the batch sample was 10-fold higher (1.03 ± 0.02 g/L) than for the fed-batch sample (0.11 ± 0.002 g/L), thereby lowering the probability of the uptake of labeled glucose from the quenching solution. In both cases, the glucose uptake was not stopped instantaneously when the cells were transferred into the quenching solution. The relative fraction of the ^{13}C-labeled metabolites, however, was below 50 %. The complete quenching took less than a second. Consequently, the metabolism was stopped before the cells, having a specific glucose uptake rate of 15.2 mmol/(g·h) and 2.7 mmol/(g·h) for batch and fed-batch cultivation (derived from monitored process data, Appendix: Figure 9.1), respectively, could consume the available glucose in the sample. In both cases, the present amount of glucose was high enough to suffice for several minutes of growth. Next, the AEC of the samples was determined, similarly with and without the addition of glucose into the quenching solution, respectively. The obtained values (Table 5.3) were well within the range of viable cells defined by Chapman et al. (1971).

Table 5.3: Adenylate Energy Charge (AEC) of *E. coli* K12 DSM 2670 cultures grown in a bioreactor using defined minimal medium with unlabeled glucose as the sole carbon source. Measurements (n=2) were performed for cultures growing in batch and fed-batch mode, respectively. The quenching solution, which was used to stop cell metabolism during the sampling process, was spiked with 1 g/L glucose (QS + Glc) or lacked additional glucose (QS – Glc).

Mode	AEC [-]	
	QS + Glc	QS - Glc
batch	0.77 ± 0.003	0.88 ± 0.003
fed-batch	0.91 ± 0.000	0.79 ± 0,002

In a next step, two extraction protocols were tested and compared. In short, the extraction efficacy of boiling ethanol was compared with that of cold acidic acetonitrile methanol. As test scenario, *C. glutamicum* was cultivated in baffled shake flasks (n=3) using defined minimal medium with glucose as the sole carbon source. The culture was sampled during exponential growth (at a CDW of about 3.5 g/L), extracted according to the two protocols and the extracts were analyzed by LC-MS/MS. The extraction efficacy was deduced from the obtained metabolite pools (Table 5.4) obtained by application of both approaches. Obviously, the softer condition produced higher errors. This could be a consequence of residual metabolic activity and precipitation of polar compounds.

Table 5.4: Comparison of intracellular levels of metabolites of the central carbon metabolism obtained by extractions (n=2) using Acidic Acetonitrile Methanol (AAM) mixture and Boiling Ethanol (BE), respectively. *C. glutamicum* ATCC 13032 was grown in baffled shake flasks containing minimal medium using glucose as sole carbon source. Samples were taken during exponential growth at an OD_{660} of 6.

	AAM			BE		
Compound	Conc. [µmol/g_{CDW}]			Conc. [µmol/g_{CDW}]		
G6P	19.6	±	3.4	24.4	±	3.2
F6P	1.2	±	0.2	1.8	±	0.3
FBP	118.8	±	74.2	189.4	±	44.1
DHAP	0.7	±	0.1	0.8	±	0.1
GAP	0.5	±	0.2	0.5	±	0.1
3PG	10.8	±	13.5	36.8	±	19.8
2PG	0.8	±	2.1	2.2	±	0.8
PEP	1.2	±	0.5	1.6	±	1.5
6PG	2.3	±	0.8	3.0	±	1.3
X5P/RIBU5P	0.1	±	0.0	0.0	±	0.1
R5P	0.0	±	0.0	0.0	±	0.1
S7P	3.3	±	0.8	6.8	±	1.9
AcCoA	0.2	±	0.1	0.2	±	0.1
α-KG	0.7	±	2.2	0.9	±	0.2
SucCoA	10.9	±	2.9	11.5	±	2.6
NAD	0.3	±	0.1	0.5	±	0.1
NADH	19.7	±	8.2	14.9	±	4.3
NADP	0.9	±	0.4	2.3	±	0.3
NADPH	9.0	±	2.9	6.7	±	1.5
FAD	0.1	±	0.1	0.3	±	0.0
AMP	3.1	±	1.9	0.4	±	0.0
ADP	26.2	±	10.6	58.9	±	7.7
ATP	45.3	±	14.1	97.7	±	19.4

Figure 5.10 depicts the relative extraction efficacy. For the vast majority of metabolites, higher concentrations were achieved using boiling ethanol, suggesting this extraction method as preferable. A possible reason for the reduced extraction efficacy might be the high amount of solvent in the acetic-acetonitrile-methanol mixture, which might lead to precipitation of polar compounds during the extraction (Canelas et al. 2009). Exceptions were phosphorylated pentoses, NADH, NADPH and AMP, which seemed to be better extracted by acidic acetonitrile methanol. The apparently higher concentrations of NADH, NADPH and AMP while using softer extraction conditions, could be a consequence of the slower inactivation of enzymes in cold acetic-acetonitrile-methanol in contrast to the almost sudden degradation of enzymes in boiling ethanol. Thus, boiling ethanol was selected as most suitable extraction liquid for further studies.

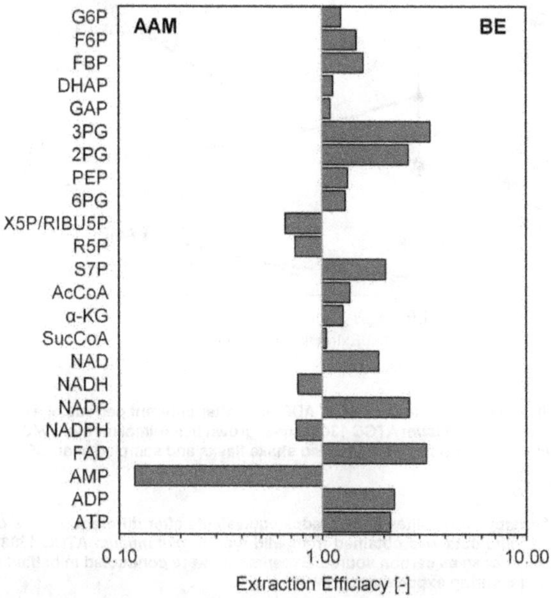

Figure 5.10: Comparison of the relative extraction efficacy of cooled Acidic Acetonitrile Methanol (AAM) mixture and Boiling Ethanol (BE), using *C. glutamicum* ATCC 13032 as the representative strain. The microorganism was grown in baffled shake flasks containing minimal medium with glucose as carbon source.

To enhance the extraction efficacy of boiling ethanol for AMP, NADH and NADPH, the extraction time was varied (Figure 5.11). For this purpose, *C. glutamicum* was cultivated in shake flasks (n=3) using minimal medium with glucose as carbon source. The cultures were sampled during the exponential growth phase and extracted with boiling ethanol for 1, 2.5 and 4 min, respectively. With increasing extraction periods, the concentration of ATP decreased while AMP and ADP concentrations increased. This seems to be a consequence of dephosphorylation of ATP to ADP and further on to AMP due to the prolonged heat treatment. This effect was particularly visible for an extraction time of 4 minutes. However, the short extraction period of 1 min caused substantially high variation between replicates, thus the medium extraction period of 2.5 min was chosen to provide a reproducible extraction process. A similar behavior was observed for the NADPH/NADP ratio (Table 5.5). With increasing extraction time, the ratio of NADPH/NADP decreased by more than 50%. However, the ratio of NADH/NAD remained stable.

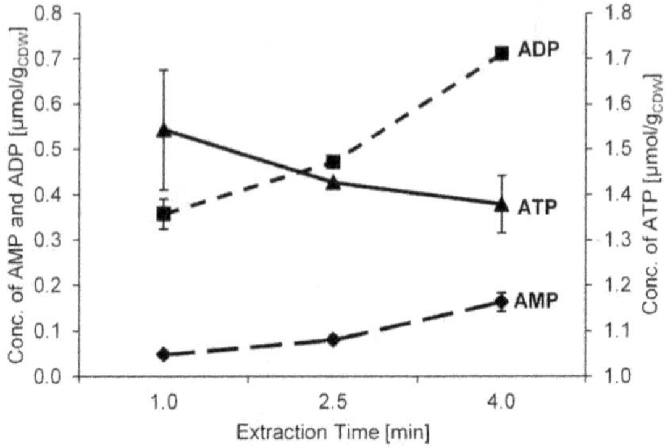

Figure 5.11: Intracellular concentrations of AMP, ADP, ATP after different periods of extraction with boiling ethanol, obtained from *C. glutamicum* ATCC 13032 (n=3) grown in minimal medium with glucose as carbon source. Experiments were conducted using baffled shake flasks and sampled at an OD_{660} of 6 during exponential growth.

Table 5.5: Ratios of energy metabolites and of redox equivalents after different periods of extraction with boiling ethanol. Metabolite data was obtained from wild type *C. glutamicum* ATCC 13032 (n=3) grown in minimal medium with glucose as carbon source. Experiments were conducted in baffled shake flasks and sampled at an OD_{660} of 6 during exponential growth.

Time [min]	AEC [-]	NADH/NAD [-]	NADPH/NADP [-]
1.0	0.88 ± 0.09	0.03 ± 0.004	1.08 ± 0.12
2.5	0.84 ± 0.01	0.02 ± 0.004	0.87 ± 0.03
4.0	0.77 ± 0.01	0.03 ± 0.005	0.41 ± 0.05

Summarizing it can be stated, that the extraction of prokaryotes with boiling ethanol showed higher extraction efficacy and better reproducibility than the acetic-acetonitrile-methanol method. Consequently, all further extractions were performed using boiling ethanol as the extraction method. Furthermore, it could be shown, that an extraction time of 2.5 minutes resulted in reproducible results for energy metabolites and redox equivalents and only minor changes in metabolite levels.

5.1.7 Validation by thermodynamic constraints

The new protocol was next used to quantify intracellular metabolites of two model organisms: *C. glutamicum* ATCC 13032 and *E. coli* K12 DSM 2670. Both organisms were cultivated in shake flasks (n=3) using defined media with glucose as carbon source. Both cultures were harvested during exponential growth phase. At the time point of sampling more than 1 g/L glucose was still left in the medium (data not shown) so that glucose starvation during the sampling process was avoided. The samples were quenched, extracted and analyzed by LC-MS/MS as described above (Table 5.6).

Table 5.6: Intracellular concentrations of 29 metabolites from central carbon metabolism for the strains *C. glutamicum* ATCC 13032 and *E. coli* K12 DSM 2670 (n=3). Both strains were grown in baffled shake flasks using minimal medium with glucose as carbon source. Strains were sampled during exponential growth. The ^{13}C-labeled extract (100 µL) was added to the biological sample as an internal standard, before extraction of metabolites by boiling ethanol (t=2.5 min) was performed.

Compound	*C. glutamicum* [nmol/g_{CDW}]			*E. coli* [nmol/g_{CDW}]		
PYR	195	±	96	2044	±	2838
FUM	4611	±	11351	880	±	63
SUC	254	±	1702	761	±	69
MAL	1379	±	9040	4265	±	753
α-KG	7700	±	10296	3474	±	371
PEP	116	±	66	1368	±	119
DHAP	5803	±	1035	2673	±	378
GAP	45	±	131	504	±	191
2PG	61	±	46	439	±	37
3PG	3114	±	1602	2765	±	149
CIT	5932	±	6212	988	±	1808
ISOCIT	671	±	446		< LOD	
R5P	1000	±	114	525	±	51
RIBU5P	748	±	163	108	±	29
G6P	4248	±	434	4257	±	150
F6P	790	±	283	733	±	70
6PG	1131	±	577	425	±	5
S7P	1858	±	677	840	±	56
FBP	38158	±	5578	5959	±	383
AMP	1687	±	716	691	±	45
AcCoA	130	±	11	778	±	17
ADP	2453	±	460	2562	±	27
SucCoA	731	±	245	6	±	5
ATP	5434	±	1481	7268	±	245
NAD	1001	±	59	1588	±	49
NADH	302	±	36	421	±	41
NADP	540	±	43	575	±	7
NADPH	569	±	91	608	±	47
FAD	211	±	21	166	±	10

Obviously, the quantitation method was sensitive enough to detect the majority of metabolites from central carbon metabolism for both microorganisms. Significantly higher errors for organic acids probably result from the fact, that these compounds were also found outside the cells, which interferes with the differential estimation (Zakhartsev et al. 2015). A first inspection of the data revealed an energy level of 0.70 (*C. glutamicum*) and 0.81 (*E. coli*), indicating physiologically viable cells (Bolten et al. 2007) and a suitable analytical approach.

Table 5.7: Optimal reaction free energies for biochemical reactions of the central carbon metabolism under physiological conditions (Li et al. 2011)

Pathway	Reaction	ΔG°' [kJ/mol]
EMP	G6P → F6P	2.78
	F6P → FBP	-15.82
	FBP → GAP + DHAP	24.64
	DHAP → GAP	7.57
	GAP → 3PG	-19.00
	3PG → 2PG	6.35
	2PG → PEP	-4.47
	PEP → PYR	-27.18
	PYR → AcCoA	-39.26
PP	G6P → 6PG	-21.89
	6PG → RIBU5P	1.23
	2 6PG → GAP + S7P	-0.62
	GAP + S7P → F6P + GAP	-6.08
TCA cycle	MAL → AcCoA+ OA → CIT	-8.56
	CIT → ISOCIT	0.92
	ISOCIT → AKG	-3.29
	AKG → SUC	-43.18
	SUC → FUM	-0.59
	FUM → MAL	-3.52

Next, a fundamental thermodynamic evaluation of all metabolite data was conducted. For both organisms the direction of pathway fluxes during growth on glucose is well known from direct ^{13}C metabolic flux analysis (Wittmann & Heinzle 2002; Hiller 2006). Using the free Gibbs energies for each biochemical reaction (Table 5.7) (Cornish-Bowden 1981), the measured metabolite data were used to prove if they provide driving forces, which support a reaction in agreement with the known metabolic fluxes.

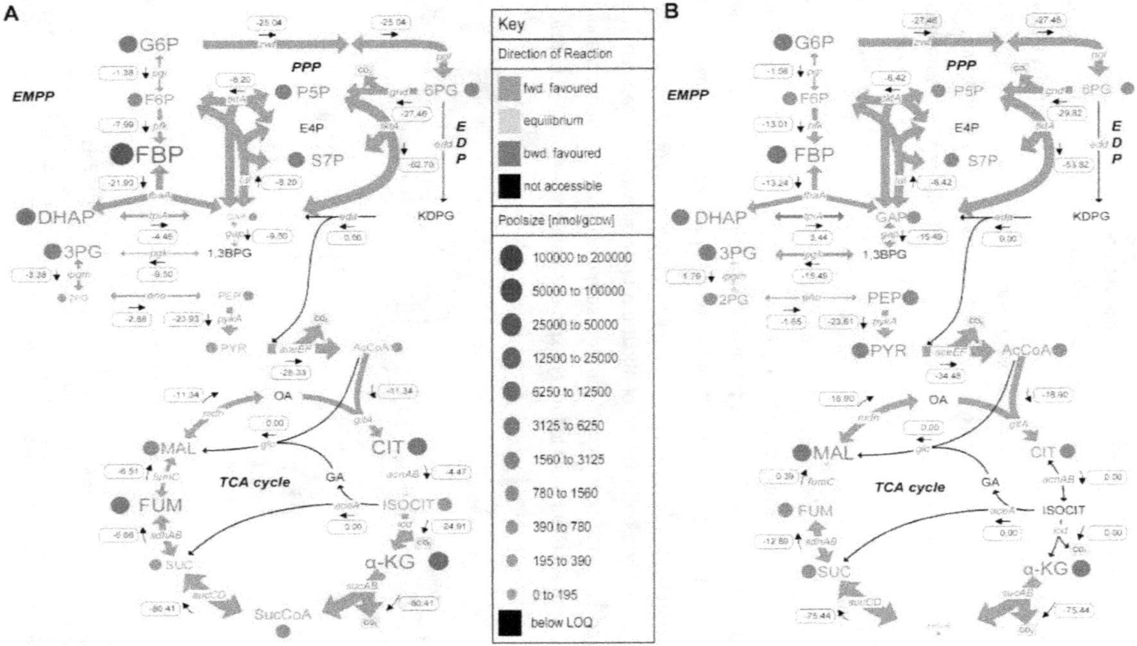

Figure 5.12: Results of the evaluation of the metabolic datasets of *C. glutamicum* ATCC 13032 (A) and *E. coli* K12 DSM 2670 (B) by thermodynamic constraints and driving force, which were computed from the measured metabolite concentrations. Color and size of the letters of the metabolites and the adjacent circles represent the measured pool size. The color of the connecting arrows indicates in which direction the reaction would proceed for the measured metabolite concentrations, while the adjacent values represent the computed Gibbs energies for the corresponding reaction. Negative Gibbs energies support a forward directed reaction, positive energies result in a back reaction. Both microorganisms were grown in baffled shake flasks containing minimal medium with glucose as the carbon source. Cultures were sampled during exponential growth at an optical density of 6. GA and KDPG could not be found in the metabolic samples due to the inactivity of the ED pathway and the GA cycle. Consequently, no driving forces were calculated for reactions using these two metabolites.

Here, analysis of the flux direction was performed by calculation of driving forces and Gibbs energies for every reaction as described earlier (see chapter 4.8). The results are shown in Figure 5.12. All reactions proven thermodynamically feasible on basis of the metabolite levels are given in green, whereby the metabolite levels themselves are visualized as circles. Most important, the evaluation by thermodynamics proved that the obtained intracellular concentrations were at reasonable levels (Figure 5.12). The metabolite pools were far from the equilibrium, thus providing a driving force to push metabolism. This is indicated by the resulting negative values for the Gibbs energies. Consequently, the newly developed metabolomics approach provided datasets of high quality for Gram-negative and Gram-positive microorganisms.

5.2 Quantitative Analysis of the Energy Metabolism of *E. coli*

5.2.1 Dynamics of the Energy Level during Carbon Deprivation

The AEC of microorganisms, especially *E. coli*, is buffered at a value around 0.8. Its dynamics are correlated with growth conditions, such as the availability of an energy source or oxygen in the surrounding medium (Chapman et al. 1971). It is generally assumed that growing cells exhibit an AEC between 0.7-0.9 (Atkinson 1968). Early experiments, however, demonstrate that energy levels transiently drop to values of about 0.5 upon starvation (Chapman et al. 1971). The developed approach now allowed the precise quantitation of underlying energy metabolites which was an excellent basis to further explore the energy dynamics of *E. coli*.

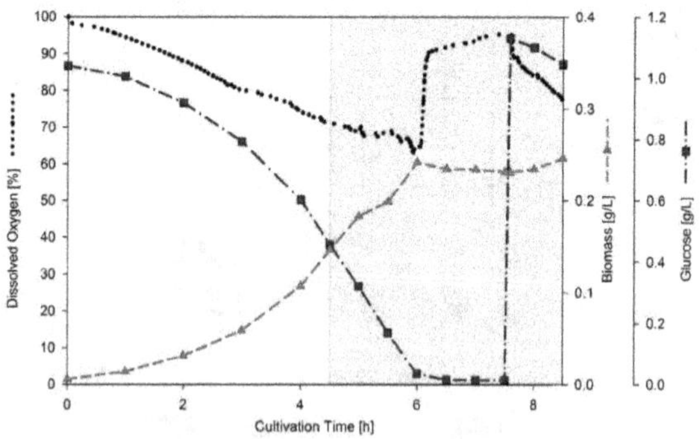

Figure 5.13: Growth and oxygen consumption of *E. coli* K12 DSM 2670 during periods of glucose excess and depletion. *E. coli* grew in baffled shake flasks filled with minimal medium (Chapman et al., 1971), containing glucose as sole carbon source. The initial amount of glucose (1 g/L) was depleted after 6 hours resulting in starvation of *E. coli*. After 1.5 hours of starvation, the glucose pool was repleted by external addition of a glucose pulse. A more detailed look at the starvation period (grey marked area) is given in Figure 5.14.

Consequently, *E. coli* was cultivated under glucose excess conditions (Chapman et al. 1971) in a lab scale bioreactor. During the cultivation, pH value (pH 7±0.05), oxygen saturation (>30%) and temperature (37±0.2 °C) were tightly controlled. The cultivation profile is depicted in Figure 5.13.

Cells grew exponentially (μ=0.69 h^{-1}) during the first 6 hours until the available glucose was depleted. This caused an immediate cessation of growth. After 1.5 hours of starvation a glucose pulse was added, thereby restarting growth (μ=0.07 h^{-1}). The level of dissolved oxygen declined constantly during both growth phases, but showed a rapid increase back to 95 % of saturation during the starvation period. Metabolite samples (n=2) were around the glucose starvation period to determinate intracellular and extracellular concentrations of AMP, ADP and ATP as well as the resulting adenylate energy charge (Figure 5.14).

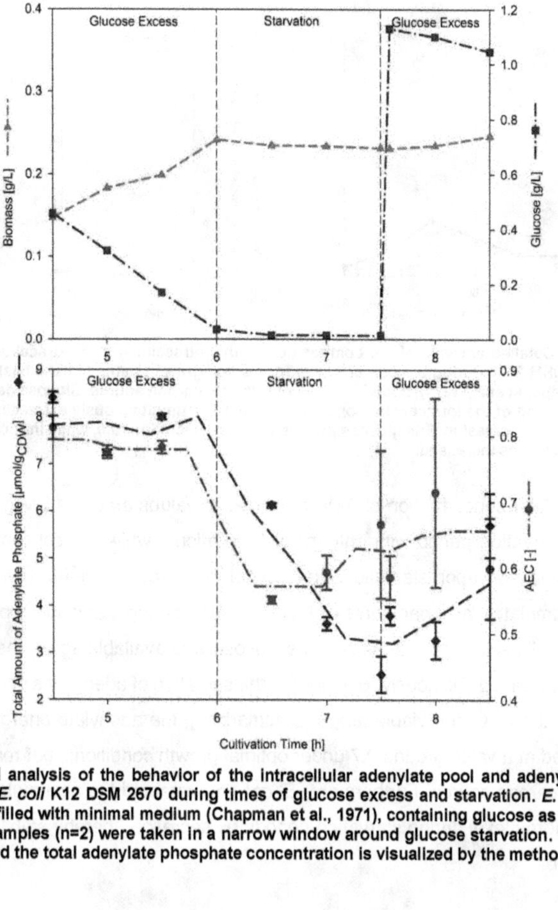

Figure 5.14: Detailed analysis of the behavior of the intracellular adenylate pool and adenylate energy charge exhibited by *E. coli* K12 DSM 2670 during times of glucose excess and starvation. *E. coli* grew in baffled shake flasks filled with minimal medium (Chapman et al., 1971), containing glucose as sole carbon source. Metabolite samples (n=2) were taken in a narrow window around glucose starvation. The trend of the energy charge and the total adenylate phosphate concentration is visualized by the method of running average.

During the first 6 hours of the cultivation the energy level was within the expected range for growing cells of 0.8 (Atkinson & Walton 1967). As soon as the glucose concentration became limiting, the AEC dropped to a value of about 0.5, where it remained during the starvation period. The carbon replenishment after 1.5 hours of starvation induced the recovery of the AEC. The total concentration of all adenylate phosphates (AMP, ADP and ATP) showed a similar behavior. While it was stable during growth, a rapid decline by more than 50% was observed upon starvation. As soon as the conditions supported further growth, the overall concentration of the adenylate phosphates increased again. These findings were in good agreement with the previous studies of starving *E. coli* cells by Chapman et al. (1971). Beyond these data, a closer look on the individual intracellular and extracellular concentrations of AMP, ADP and ATP was taken (Figure 5.15).

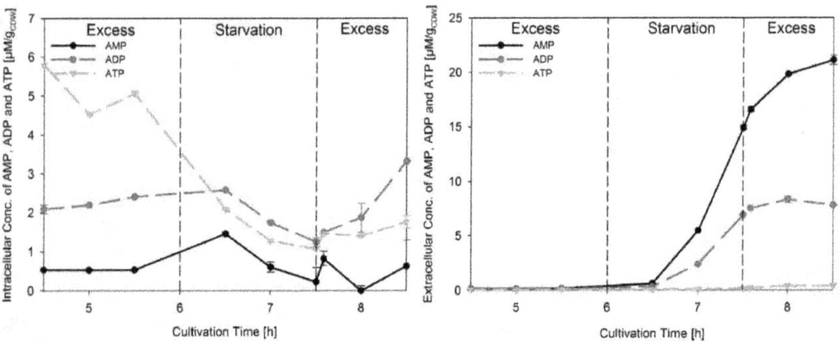

Figure 5.15: Detailed analysis of the composition of the intracellular and extracellular adenylate pool of *E. coli* K12 DSM 2670 during periods of starvation. *E. coli* grew in baffled shake flasks filled with minimal medium (Chapman et al., 1971), containing glucose as sole carbon source. Glucose depletion was reflected by a sudden drop of the intracellular concentration of ATP, simultaneously extracellular amounts of AMP and ADP started increasing. The glucose pulse slowed the secretion rate, while intracellular energy metabolite concentrations increased.

The intracellular concentration of AMP remained at values around 1 µM/g_{CDW} throughout the complete starvation period with only minor fluctuations, while the concentration of ATP decreased immensely upon starvation. The level of ADP dropped slightly delayed. Furthermore, *E. coli* accumulated larger amounts of AMP and ADP during starvation, while ATP was effectively retained inside the cells. As soon as glucose was available again, the secretion of AMP and ADP decreased. Obviously, *E. coli* used the secretion of adenylates of lower energy levels to maintain the AEC in a viable range. Summarizing, the adenylate energy charge of *E. coli* was buffered at a value around 0.7 under optimal growth conditions, but reacted to the depletion of the carbon source. Furthermore, *E. coli* counteracted the accumulation of adenylates with low energy level during starvation periods by secretion, thereby stabilizing the AEC in a viable range.

5.2.2 Dynamics of Energy Metabolism during Chemostat Experiments

Next, the adenylate phosphate levels were assessed for *E. coli* K12 DSM 2670 growing in chemostat cultures at different dilution rates (D). Cultivation profiles obtained for the tested dilution rates of D=0.1 h^{-1}, D=0.2 h^{-1} and D=0.4 h^{-1} are depicted in Figure 5.16 A, B and C, respectively. The cultivations were sampled (n=3) after five residence times.

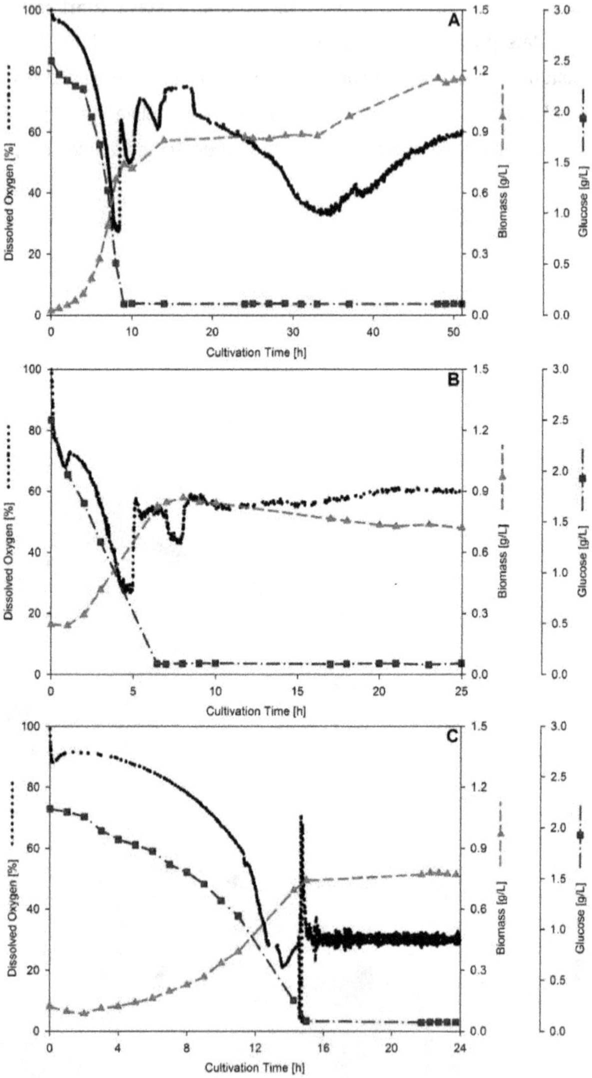

Figure 5.16: Cultivation profiles of *E. coli* K12 DSM 2670 grown in chemostat at differing dilution rates D=0.1 h^{-1} (A), D=0.2 h^{-1} (B) and D=0.4 h^{-1} (C) in a 1L-bioreactor filled with 0.4 L minimal medium, containing 2.5 g/L glucose as carbon source. The minimal medium was continuously pumped into the reactor at a preset rate, while the effluent was drained by a slightly higher pump rate, thereby keeping the volume constant. Dissolved oxygen was always maintained above 30 % by control of stirrer speed and aeration rate. During the cultivation at a dilution rate of D=0.1 h^{-1} the drain pipe slid slightly into the reactor. The decrease in cultivation volume resulted in a higher glucose concentration in the reaction vessel and consequently in an enhanced biomass formation and a drop of oxygen saturation (Figure 5.16A). However, the culture was sampled as the biomass formation had adapted to the changed conditions and no further biomass growth had been monitored for 4 consecutive hours.

Table 5.8: Intracellular concentrations of AMP, ADP, ATP of *E. coli* K12 DSM 2670 growing in minimal medium (0.4 L) with glucose as the carbon source. Cells were grown in 1L-bioreactors operated in chemostat at differing dilution rates, which were sampled after 5 residence times. All measurements were performed as triplicates (n=3).

Dilution Rate D	Metabolite Concentration [µM]		
[h^{-1}]	AMP	ADP	ATP
0.1	0.44 ± 0.07	1.22 ± 0.08	1.70 ± 0.12
0.2	0.36 ± 0.02	2.73 ± 1.21	7.30 ± 2.00
0.4	0.57 ± 0.20	2.88 ± 0.79	6.08 ± 1.67

The energy charge (AEC) and the overall adenylate concentration, derived from the corresponding AMP, ADP and ATP level (Table 5.8), are shown in Figure 5.17. For all dilution rates, the energy charge was well within range of growing cells (Figure 5.17A). The concentration of the adenylate phosphates, however, responded to the change in growth rate. At the lowest growth rate of D=0.1 h^{-1}, the overall concentration of adenylate phosphates was only 30 % of the level observed at higher growth rates. An explanation of these results could be the decreased metabolic activity at the lower growth rate. From an economical point of view, it seems logical, that a cell produces only as much adenylate phosphates, as are needed to store the energy produced under the given conditions. These correlations of AEC, adenylate pool and dilution rate are in good agreement with other prokaryotes (Swedes et al. 1975; Marriott et al. 1981) and fungi (Vu-Trong & Gray 1982).

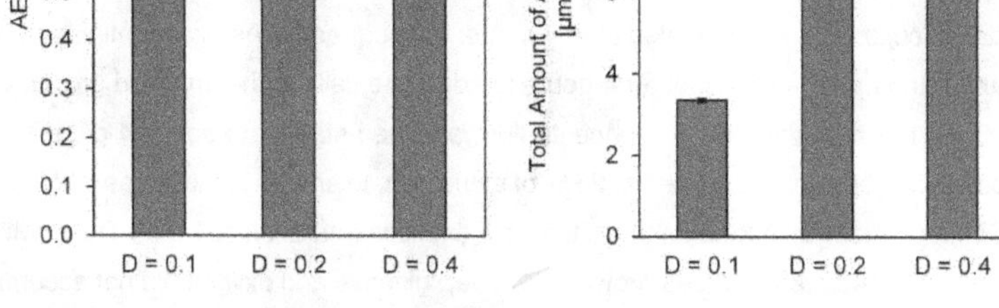

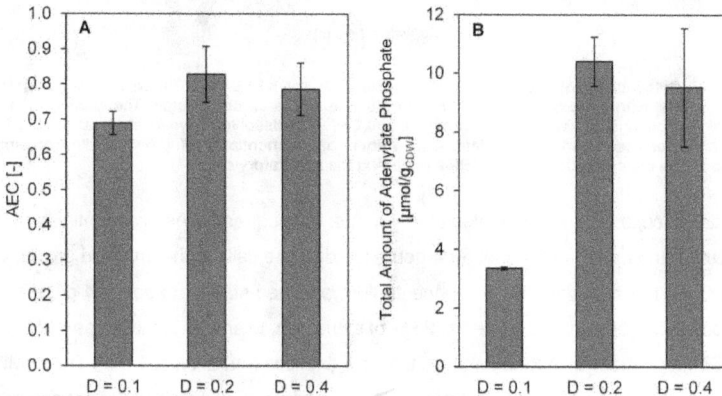

Figure 5.17: Influence of dilution rates on the adenylate energy charge (A) and total adenylate phosphate pool (B) of *E. coli* K12 DSM 2670 growing in minimal medium (0.4 L) with glucose as carbon source. Cells were grown in 1L-bioreactors operated in chemostat at differing dilution rates, which were sampled after 5 residence times. All measurements were performed as triplicates (n=3).

5.2.3 Inhibition of the Respiratory Chain and Impact on the Energy Charge

Dinitrophenol was next used to induce an artificial distortion of the energy metabolism (Dietzler et al. 1979; Dills & Klaassen 1986), thereby enabling the investigation of effects on the intracellular energy level. The compound functions as uncoupling agent of the respiratory chain and the phosphorylation system, resulting in reduced ATP synthesis and thus an imbalanced energy charge. In short, *E. coli* was cultivated in a fed-batch process using an exponential feed profile, which supported a constant growth rate of μ=0.2 h^{-1} (Figure 5.18).

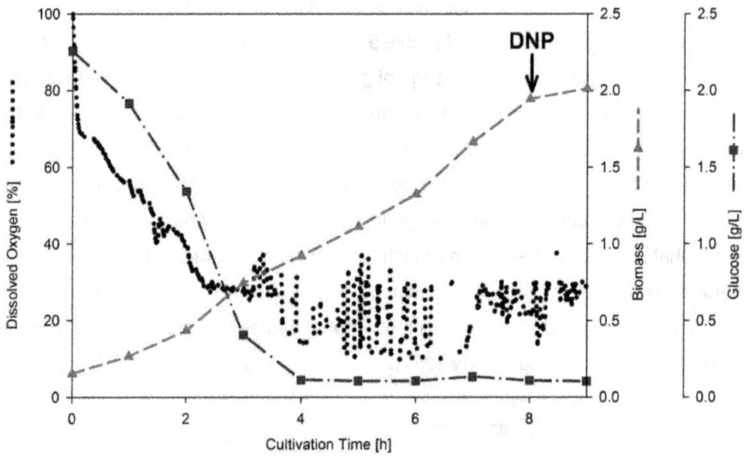

Figure 5.18: Cultivation profile of a fed-batch process of *E. coli* K12 DSM 2670. Cells were grown in a 1L-bioreactor filled with 0.4 L defined medium with glucose as the carbon source. The applied feed profile supported exponential growth at a growth rate of μ=0.2 h^{-1}. The dissolved oxygen level was kept at 25 % by control of stirrer speed and aeration rate. After 4 hours of exponential feeding (t=8 h), dinitrophenol was added to a final concentration of 1 mM, thus inhibiting the respiratory chain.

The initial glucose level was depleted after 4 hours. Subsequently, the exponential feed profile was started and operated for another 4 hours to adapt the cells to the imposed growth conditions. During this time the glucose concentration remained stable at about 0.1 g/L. The level of dissolved oxygen was maintained at 25 % of saturation, to ensure sufficient aeration (Figure 5.18). Dinitrophenol was added to the culture to a final concentration of 1 mM. The addition of dinitrophenol immediately reduced growth. However, glucose and oxygen did not accumulate, indicating that the specific glucose and oxygen uptake rates remained constant (Figure 5.18). Samples for the measurement of the intracellular adenylate phosphate pools (n=3) were taken directly before the addition of dinitrophenol and one hour later (Figure 5.19 A and B). Obviously, the uncoupling of the respiratory chain was accompanied by a substantial decrease of the energy level, while the total adenylate pool decreased only little.

Table 5.9: Intracellular concentrations of AMP, ADP and ATP obtained from exponentially growing *E. coli* K12 DSM 2670 (μ=0.2 h^{-1}) prior to and after the addition of 1 mM dinitrophenol to the culture. Cells were grown in a 1L-bioreactor filled with 0.4 L defined medium with glucose as the carbon source. Measurements were performed as triplicates (n=3).

DNP	Metabolite Concentration [µM]								
	AMP			ADP			ATP		
-	0.80	±	0.27	3.24	±	0.60	5.78	±	0.62
+	2.30	±	0.99	3.64	±	0.66	2.72	±	0.26

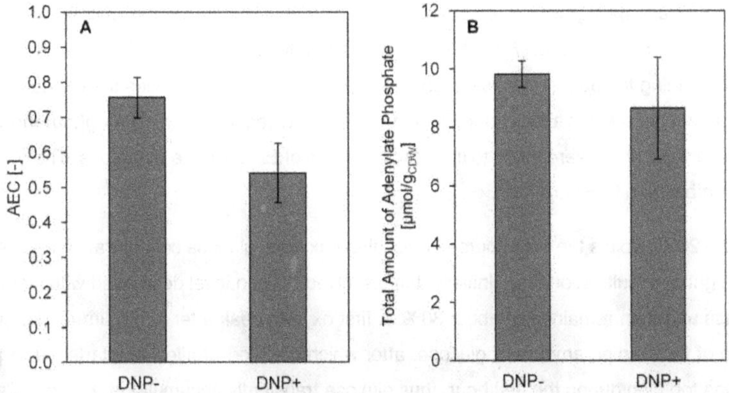

Figure 5.19: Effect of dinitrophenol (1 mM) on the energy metabolism of *E. coli* K12 DSM 2670, i.e. adenylate energy charge (A) and the total amount of intracellular adenylate phosphates (B) of *E. coli* (n=3). Cells were grown in a 1L-bioreactor filled with 0.4 L defined medium with glucose as the carbon source.

It is known that dinitrophenol leads to a collapse of the proton gradient, thereby disabling the generation of ATP via the respiratory chain. Thus, the major reaction to generate ATP was no longer functional after the addition of the uncoupling agent. This is reflected by the lowered AEC (Figure 5.19A). The reduced AEC value leads to higher activities of regulatory enzymes, such as phosphofructokinase, and consequently to an upregulated flux through glycolysis to compensate for the inefficient energy generation (Atkinson & Walton 1967; Wittmann et al. 2007). These results point out that *E. coli* can still grow at low energy levels as long as a carbon source is available. The observed low AEC value further indicates an imbalance of the energetic state, but seems not ultimately linked to complete cessation of growth as previously suggested (Chapman et al. 1971). Hence, such low AEC values are fully conclusive for corrupted datasets, but could also be the consequence of an imbalanced energy metabolism. Extreme growth conditions would be expected to generate such imbalances. As example, energy demanding genetic modifications or biosynthetic demands might lead to imbalances between energy demand and supply, thereby lowering the AEC (Wittmann et al. 2007). Similarly, it was reported that *Penicillium simplicissimum*, which were grown in glucose-limited chemostat cultures, respond to sudden glucose excess with short termed fluctuations of adenylate phosphate concentrations and AEC (Ganzera et al. 2006). As shown above, the AEC seems to be more variable than previously thought.

5.2.4 Adenylate Energy Charge during Fed-batch Cultivations

Following the previous results, an experiment was conducted to investigate the effects of limitation and excess conditions, regarding glucose and oxygen supply, in more detail. For this purpose, *E. coli* was grown in fed-batch mode using an exponential feed profile, which supported a growth rate of μ=0.17 h^{-1}. At a certain point of time, a glucose pulse was added to the reactor, leading to sudden glucose excess. During both cultivation modes the oxygen regulation was suspended for a two-hour period to induce oxygen limitation. Throughout the whole experiment, samples were taken to track the response of *E. coli* to the limitations. The obtained data can be seen in Figure 5.20.

Figure 5.20 A depicts the time course of dissolved oxygen, glucose concentration and cellular dry weight over cultivation time. Initially, the dissolved oxygen level decreased with increasing biomass and then remained at about 30%. A first oxygen peak after 6.5 hours marks the depletion of the starting amount of glucose, after which the feed profile was started. The pump rate was too high during the first hour, thus glucose transiently accumulated. However, a second oxygen peak at t=9.3 h marked the point from which the glucose-limitation controlled the growth rate. The growth rate was maintained at μ=0.17 h^{-1} by the feed. The glucose pulse at t=17.5 h led to a glucose concentration of about 10 g/L, thereby ending the carbon limitation and resulting in an increased growth rate. Furthermore, it can be seen that the suspended dissolved oxygen control resulted in an immediate drop of the oxygen level to 0 % during both two-hour periods of oxygen limitation (t=13-15 h and t=20-22 h). Reactivation of the oxygen regulation led to an immediate increase to the preset saturation of 30 % dissolved oxygen in the culture broth. Obviously, ATP was the predominant species of the adenylates during the batch and the early feed phase (Figure 5.20B). The intracellular levels of AMP, ADP and ATP increase proportionally with the biomass during these first 6 hours of cultivation. However, a significant drop of the intracellular ATP concentration was monitored, as soon as the feed was started. During the following hours, the ATP level decreased constantly until it reached a minimum of about 0.6 $\mu mol/g_{CDW}$ after t=14 h. The ATP level was then maintained at this value. The amounts of AMP and ADP, on the other hand, continuously increased and peaked after 13 hours of cultivation, before the first period of oxygen limitation was started. Both periods of oxygen limitation were accompanied by a sudden decrease in the concentration of AMP and ADP, respectively. However, the concentrations of both metabolites increased immediately,

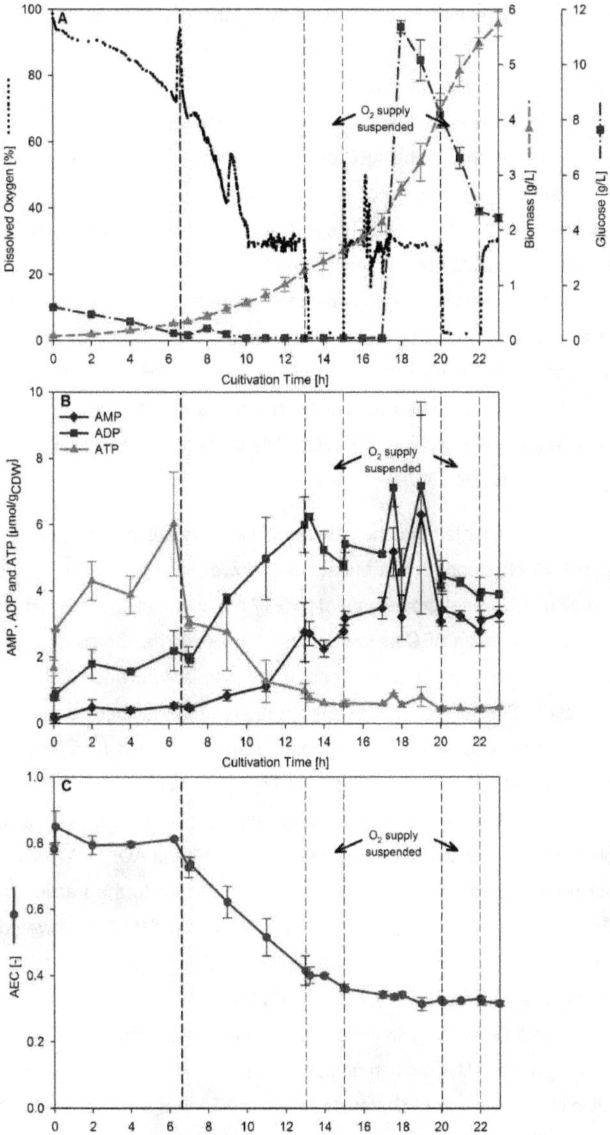

Figure 5.20: (A) Time profile of dissolved oxygen level, glucose concentration and biomass formation of *E. coli* K12 DSM 2670 during a fed-batch cultivation. Cells were grown in a 1L-bioreactor filled with 0.4 L minimal medium with glucose as the carbon source. After 6 hours, the initial glucose was depleted. Subsequently an exponential feed profile was started (red dashed line), which supported a specific growth rate of μ=0.17 h^{-1}. At t=17 h, a glucose pulse was added. During glucose limited growth and glucose excess the aeration was stopped for 2 hours, to induce oxygen limitation. (B) Time course of intracellular AMP, ADP and ATP. (C) Adenylate energy charge. All measurements were performed as duplicate (n=2).

when the oxygen control was re-activated. A comparison of both phases of oxygen limitation shows, that the negative effect of oxygen depletion on the A(X)P level was stronger during glucose excess (t=20 h) than during glucose limitation (t=13 h). The AEC of *E. coli* during the experiment can be seen in Figure 5.20C. The AEC correlates partly with the ATP concentration. During the batch phase, the AEC of *E. coli* was within the range for viable cells (0.7 to 0.9). The decrease of the energy charge from 0.73 to 0.34 coincided with the depletion of glucose and the initiation of the feed. The AEC reached its minimum after t=15 h of cultivation, neither the glucose excess nor the second oxygen limitation led to further negative effects. A closer look at the time slots of oxygen depletion indicates that the decrease of AEC was slowed down or even stopped by the limitation, which might be explained by the rather unchanged ATP level in contrast to the decreasing concentrations of AMP and ADP during these periods. Furthermore, the AEC remained almost unchanged after addition of the massive glucose pulse, which was unexpected, as the availability of glucose should result in a regenerating AEC and rising ATP levels.

Summarizing it can be stated that the glucose limitation resulted in a constant decrease of the AEC of *E. coli* which stopped at a substantially lower value than found by Chapman et al. (1971). However, the preset growth rate of $\mu=0.17\ h^{-1}$, controlled by the feed profile, was maintained by *E. coli* despite an AEC as low as 0.33. It is likely, that the constant glucose limitation resulted in the formation of subpopulations, which oscillate between states of starvation and ATP regeneration. Similar states of oscillation have been modeled (Cortassa 1990) and are reported for the energetic metabolism of yeast suspensions and *Klebsiella aerogenes* under starvation (Hess & Boiteux 1968; Harrison 1976). As the cells remained growing throughout the whole experiment, the fraction of starving cells possibly became higher with time, which could explain the continuously increasing levels of AMP and ADP. In addition, the fraction of cells exhibiting an optimal AEC became smaller compared to the fraction of starving cells, thereby further lowering the total AEC. According to Atkinson (1968) a lowered AEC results in reduced rates of energy utilizing enzymes, while at the same time enhancing the rates of energy regenerating enzymes. Thus, the drastically lowered AEC, exhibited by the starving cells, should actually be an advantage in the competition for the carbon source, as it accelerates the combustion of glucose. The oxygen limitation, on the other hand, led to a decrease in total concentration of adenylate phosphates, from which ATP was affected least. It is likely, that the decreased availability of oxygen led to an enhanced Crabtree effect, thereby enabling the cells to generate energy from glucose through by-product formation. The energy generated by overflow metabolism is drastically lower than the use of the respiratory chain. Thus, the cells reduced the intracellular adenylate phosphate pool to optimize their energy level. The effect of oxygen limitation was damped by the lower availability of glucose during the fed-batch process, which explains why the reduction of the adenylate phosphates is less drastic during the first phase of limitation. The observation that the energy charge of *E. coli* remained at a value of 0.33, despite the sudden glucose excess, could indicate that the cells remain in states of oscillation between energy regenerating and utilizing reactions, as proposed by the model of Cortassa (1990). Furthermore, the low energy charge could be the reason why the growth rate remained at about 0.2 h^{-1} during the glucose excess instead of increasing to μ_{max}, as a low AEC lowers the activity of anabolic enzymes.

Taken together, glucose limitation during fed-batch cultivations, led to changes in the energy charge of cells. Therefore, the question arose, to which extent *E. coli* can counteract the glucose limitation by degradation of redundant proteins and stored glycogen to maintain a high AEC as long as possible. To elucidate this in more detail, further fed-batch cultivations were necessary. However, in order to enable the LC-MS/MS method to distinguish between metabolites originating from degraded storage proteins and those from freshly added external glucose, isotopically labeled glucose was used in the feed. Furthermore, the analytical LC-MS/MS method was adapted to detect different mass isotopomers of the metabolites of the central carbon metabolism. An exception had to be made for molecules, consisting of more than seven carbon atoms, such as AMP, ADP and ATP. Here, only unlabeled and fully labeled fractions were supervised, as monitoring of all possible labeling states would have resulted in poor data quality as a consequence of a serious reduction of collected data points per minute due to longer measurement cycles.

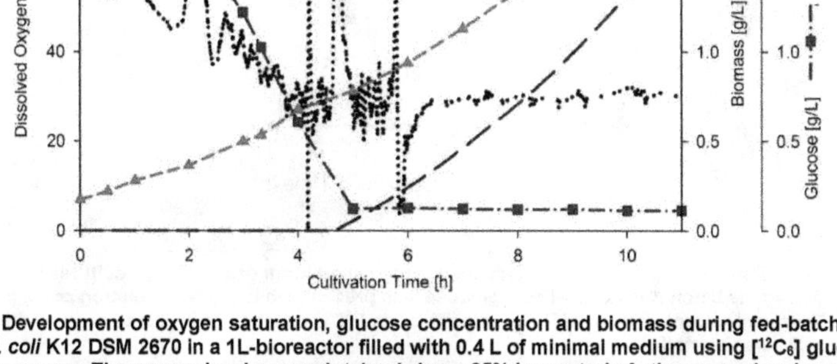

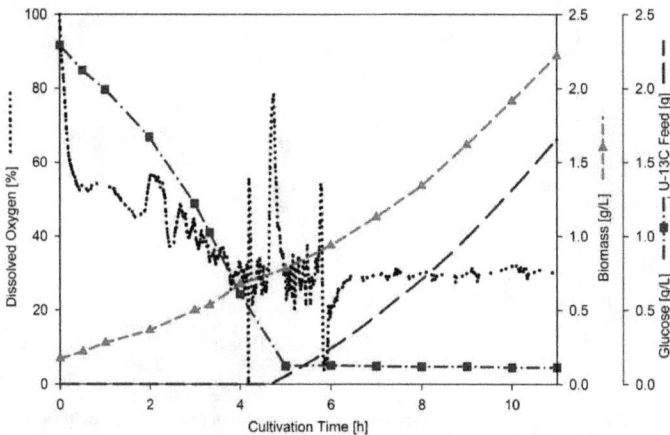

Figure 5.21: Development of oxygen saturation, glucose concentration and biomass during fed-batch cultivation of *E. coli* K12 DSM 2670 in a 1L-bioreactor filled with 0.4 L of minimal medium using [$^{12}C_6$] glucose as the carbon source. The oxygen level was maintained above 25% by control of stirrer speed and aeration rate. The feed solution contained [$^{13}C_6$] glucose to enable differentiation between carbon from storage polymers and carbon from the feed solution. The applied feed profile supported a specific growth rate of μ=0.18 h^{-1} and was started at t = 4.3 h. Additionally, the total amount of added [$^{13}C_6$] glucose is depicted.

Figure 5.21 displays the cultivation parameters of the fed-batch experiment conducted for this purpose. The starting concentration of glucose was set to 2.3 g/L to achieve a high cell density during the initial batch phase, which was needed to ensure a precise measurement of the less abundant fractions of mass isotopomers of the metabolites of the central carbon metabolism. After glucose was depleted, the exponential feed profile was started to maintain controlled growth of *E. coli*. Hereby the substrate was switched from [$^{12}C_6$] glucose to [$^{13}C_6$] glucose. *E. coli* grew with a specific growth rate of μ=0. 33 h^{-1} during batch phase, while during the feed phase, cells exhibited a lower growth rate of μ=0.18 h^{-1}. The oxygen level was kept above 30 % by control of stirrer speed and aeration rate. The total amount of added [$^{13}C_6$] glucose is illustrated in Figure 5.21. As can be seen, the glucose concentration remained at a level of 0.1 g/L during the feed phase. Thus, the medium still contained 0.1 g/L [$^{12}C_6$] glucose as the feed profile was started. Consequently, the controlled slow feeding rate of [$^{13}C_6$] glucose combined with the residual 0.1 g/L of [$^{12}C_6$] glucose in the culture broth, led to a slow replacement of [$^{12}C_6$] glucose by [$^{13}C_6$] glucose (Figure 5.22). Calculations showed that a complete labeling of the residual glucose was achieved after 2.5 h of feeding, according to the feed profile and glucose uptake rate of the cells. Obviously, the slow replacement of the residual [$^{12}C_6$] glucose by its ^{13}C isotopomer played a major role in the labeling distribution of the intracellular metabolites, which had to be considered carefully while investigating the possible backflow from protein synthesis.

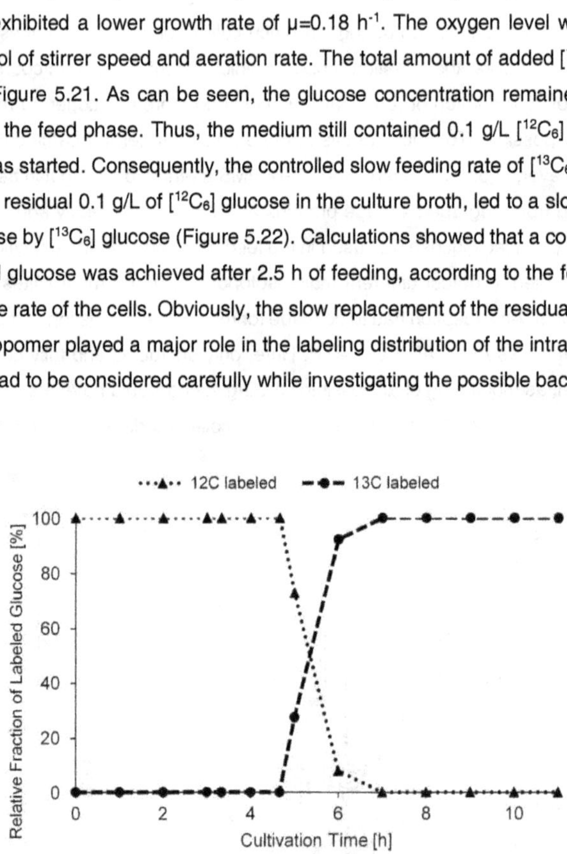

Figure 5.22: Relative fraction of labeled glucose in the supernatant of a fed-batch culture of *E. coli* K12 DSM 2670. During the batch phase only [$^{12}C_6$] glucose was present, while the feed solution contained [$^{13}C_6$] glucose. Thus, a slow replacement of [$^{12}C_6$] glucose by its [$^{13}C_6$] isotopomer began as soon as the exponential feed profile was started.

The obtained labeling patterns of all measured metabolites resemble one of the four exemplary distribution types shown in Figure 5.23. The complete set of labeling distributions can be found in the appendix (Figure 9.2 and Figure 9.3).

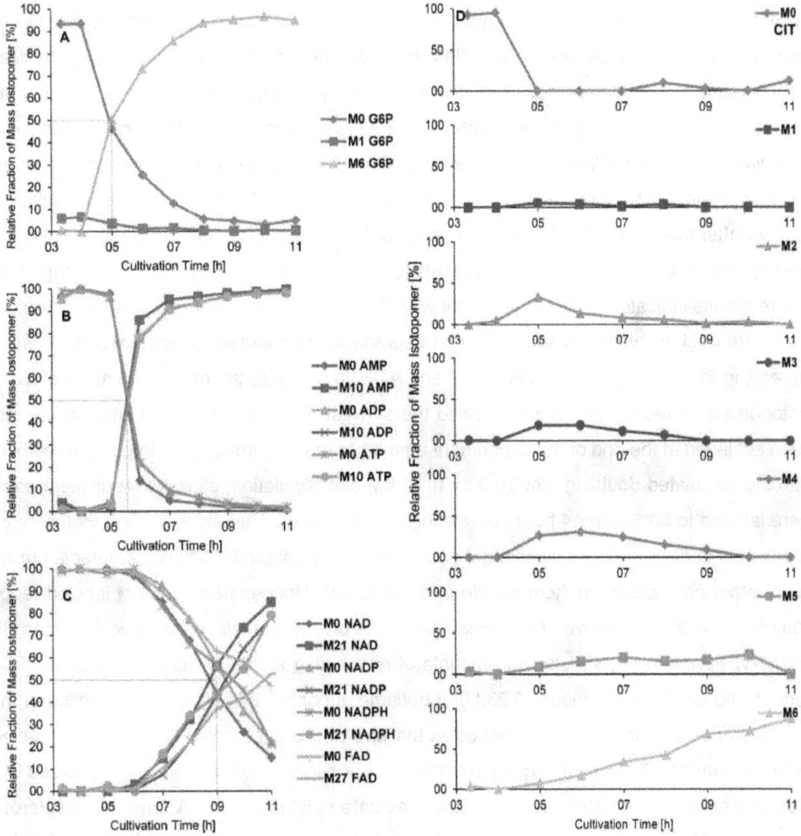

Figure 5.23: Differences in the ^{13}C labeling of metabolites from the central carbon metabolism of *E. coli* K12 DSM 2670 during the fed-batch experiment. The cells were grown in a 1L-bioreactor filled with 0.4 L minimal medium containing [$^{12}C_6$] glucose as carbon source. The feed profile supported a growth rate of 0.18 h^{-1}. The feed contained [$^{13}C_6$] glucose as the carbon source and was started as soon as the initial glucose concentration was depleted at t=4.3 h. The dotted line marks the point of time at which 50 % of the metabolites were labeled. (A) Metabolites of glycolysis and PP pathway exhibited an immediate increase of the labeled fraction of metabolites (t=4.3 h). 100% of labeling was reached at t=8 h. (B) Labeling of energy equivalents started 1 hour delayed (t=5 h), but was almost completely labeled within two hours (t=7 h). (C) Labeling of redox equivalents began 2 hours after feed start (t=6 h) and was not completed at the end of the experiment. (D) Labeling of metabolites of the TCA cycle could be seen immediately (t=4.3 h) but was not completed at the end of the experiment. Furthermore, various labeling states of the TCA cycle metabolites could be detected.

In detail, metabolites of glycolysis and PP pathway showed a strong increase of their fully labeled isotopologues, shortly after starting the feed, while only minor amounts of partially labeled isotopologues were detected (Figure 5.23 A). The labeling of the energy metabolites AMP, ADP and ATP showed a similar increase of fully labeled isotopologues. However, their dynamics were about one hour delayed, when compared to metabolites of PP pathway and glycolysis (Figure 5.23 B). This could reflect that the energy metabolites had to be synthesized from scratch, unlike metabolites of glycolysis and PP pathway, which originate directly from the labeled glucose. Furthermore, Figure 5.23 B shows that more than 50 % of the adenylate pool was fully labeled within 2 hours of feeding with [$^{13}C_6$] glucose. If the labeling pattern would have originated only from newly grown cells, a value of 50 % should have been reached 4 hours after start of the feed phase, as cells doubled every 3.85 hours (μ=0.18 h^{-1}). Consequently, the labeling rate of the energy metabolites was influenced by already existing cells. These results indicate, in good agreement with chapter 5.2.1, that *E. coli* actively regulates its adenylate pool by synthesis, excretion and degradation to maintain an optimal AEC, thereby increasing the labeling rate of AMP, ADP and ATP. In contrast, the relative fractions of mass isotopomers of redox equivalents showed the slowest response. Full [^{13}C] labeling had not been achieved at the end of the experiment after 11 hours (Figure 5.23 C). In good agreement with the calculated doubling time (t=3.85 h) of the cell population, all redox equivalent pools were labeled to 50 % after 4 hours of feeding [$^{13}C_6$] glucose. Thereby indicating that intracellular levels of the redox equivalents are not continuously adapted to fit cellular needs, but remain rather constant throughout the life cycle of *E. coli*. Hence, the fraction of labeled redox equivalents increased slower than the labeling state of energy metabolites, glycolysis and PP pathway. Furthermore, substantial amounts of all labeling states of metabolites of the TCA cycle could be detected (Figure 5.23 D). It became apparent, that a complete labeling of the TCA metabolites had not been reached at the end of the experiment. The partially labeled isotopologues are likely a consequence of the citrate synthase reaction, which synthesizes CIT from unlabeled oxaloacetate and the labeled acetate residue of AcCoA, and the anaplerotic reactions, which refill the oxaloacetate pool by coupling of unlabeled carbon dioxide to labeled PEP or PYR. Consequently, these reactions result in the observed high variation of possible labeling states of metabolites of the TCA cycle metabolites. However, these reactions did not explain the slow increase of the fraction of completely labeled metabolites of the TCA cycle. It was found that during the batch phase substantial amounts of unlabeled organic acids had been secreted into the medium, which were later used as a carbon source in the feed phase. In addition, backflows from large intracellular pools of amino acids, such as glutamate, could have decreased the labeling rate even further (Krömer et al. 2004). The exchange with unlabeled pools of carbon dioxide, extracellular organic acids and the backflow from large unlabeled intracellular pools, like glutamate, resulted in the slow increase of fully labeled CIT, ISOCIT, αKG, SUC, FUM and MAL.

The obtained relative isotopomer distributions of metabolites of glycolysis and TCA cycle indicated active degradation of storage proteins. To enable a more qualitative insight, the dynamics of the metabolite labeling, depending on the feed profile and the ratio of [$^{12}C_6$]/[$^{13}C_6$] glucose, was simulated. Shortly, it was assumed that the sequential metabolite pools behave like a set of sequential stirred tank reactors. Hence, it was possible to compute the relative fraction of the uniformly labeled mass isotopomer by applying the equation for residence time distributions in continuously stirred tank reactors (Equation 5.2), with N being the position of the metabolite in the sequence of its pathway and τ being the residence time of the same metabolite under the applied cultivation conditions.

$$F(t) = 1 - e^{-N\tau} * \sum_{1}^{N} \frac{(N*\tau)^{N-1}}{(N-1)!} \qquad \text{(Equation 5.2)}$$

The method of Wittmann et al. (2004) was utilized to calculate the required residence times for each measured metabolite. The glucose uptake rate (2.7 mmol/(g_{CDW}·h)) was derived from the experimental data and was multiplied with the flux ratios of *E. coli* grown at μ=0.125 h^{-1} (Hiller 2006). To compute the residence times (Table 5.10), the metabolite pools obtained from chemostat cultivations (D=0.1 h^{-1}) (chapter 5.2.2) were divided by the estimated flux.

Table 5.10: Intracellular fluxes of *E. coli* at a growth rate of μ=0.125 h^{-1} derived from Hiller (2006), the intracellular pool sizes of metabolites of the central carbon metabolism of *E. coli* K12 DSM 2670 at a growth rate of μ=0.1 h^{-1} (experimental data) and the resulting residence time τ.

Pathway	Metabolite	Flux [%]	Flux [µmol/(g_{CDW}*h)]	Pool Size [µmol/g_{CDW}]	Residence Time τ [min]
EMP Pathway	G6P	100.0	2695	1.46	0.032
	F6P	87.3	2353	0.09	0.002
	FBP	89.3	2407	2.44	0.061
	DHAP	89.3	2407	0.74	0.018
	3PG	167.4	4511	1.82	0.024
	2PG	167.4	4511	0.20	0.003
	PEP	167.4	4511	0.57	0.008
	AcCoA	124.3	3350	0.11	0.002
TCA cycle	CIT	98.5	2655	45.01	1.017
	ISOCIT	98.5	2655	0.03	0.001
	AKG	98.5	2655	1.92	0.043
	SUC	91.1	2455	0.33	0.008
	FUM	91.1	2455	2.15	0.053
	MAL	91.1	2455	0.35	0.009
PP Pathway	6PG	11.2	302	0.23	0.046
	P5P	11.2	302	0.18	0.036

The obtained residence times differed widely in a range from 60 ms to 1 min, as a consequence of large differences in intracellular pool sizes. The calculated residence times were applied to Equation 5.2. Here, the labeled glucose was treated as the pulse signal, which flows through sequential stirred tank reactors. Thereby it was possible to compute the relative fraction of fully labeled mass isotopomers as a function of time, which simulated the labeling patterns without backflows from proteins or glycogen. The simulated relative fractions were almost alike, despite the differing residence times. It became obvious, that the slow interchange from [$^{12}C_6$] to [$^{13}C_6$] glucose had a far stronger impact on the computed data than the actual residence times. Obviously, it took more than an hour to replace the majority of unlabeled glucose in the culture broth by the fully labeled glucose, as a consequence of the feed profile. Consequently, the impact of the residence times of less than a minute was rather small.

The comparison of simulated and experimental data revealed a great discrepancy, indicating that *E. coli* utilizes degradation of proteins or glycogen to compensate for carbon limitation (Figure 5.24). It was possible to identify three types of differences between experimental and simulated data. Characteristically the [$U^{13}C$] isotopomer fraction of the first type increased drastically during the first 20 min of the feeding process and converged towards 100 % afterwards, as the [^{12}C] labeled backflows slowed complete labeling. This type was associated with metabolites of upper glycolysis and the PP pathway and therefore influenced by additional metabolization of glycogen, histidine and eventually nucleotides. The second type was found for metabolites of the lower glycolysis and strongly resembled the first type. However, potentially additional backflow from protein degradation, which entered through the 3PG node, slowed labeling dynamics down. As a result of the additional [^{12}C] backflow, entering through the 3PG node, the labeling distribution of 3PG, 2PG and PEP became indented (Figure 5.24). Metabolites of the TCA cycle showed only a slow increase of the [$U^{13}C$] labeled isotopomers, as discussed earlier. The exchange of small intracellular and large extracellular pools of carbon dioxide and organic acids, as well as the potential backflow from glutamate and other amino acids could be the major reason of this finding. Probably supported by the backflow of carbon from proteins and storage polymers into central carbon metabolism, the AEC remained in the optimal range of 0.7 to 0.85 throughout the complete experiment. However, towards the end, the AEC decreased (AEC=0.71 at t=11h). This could indicate that the carbon reserves originating from glycogen and protein degradation were depleted. Likely, the AEC would have dropped to similar levels as obtained by the previously analyzed long term fed-batch cultivations.

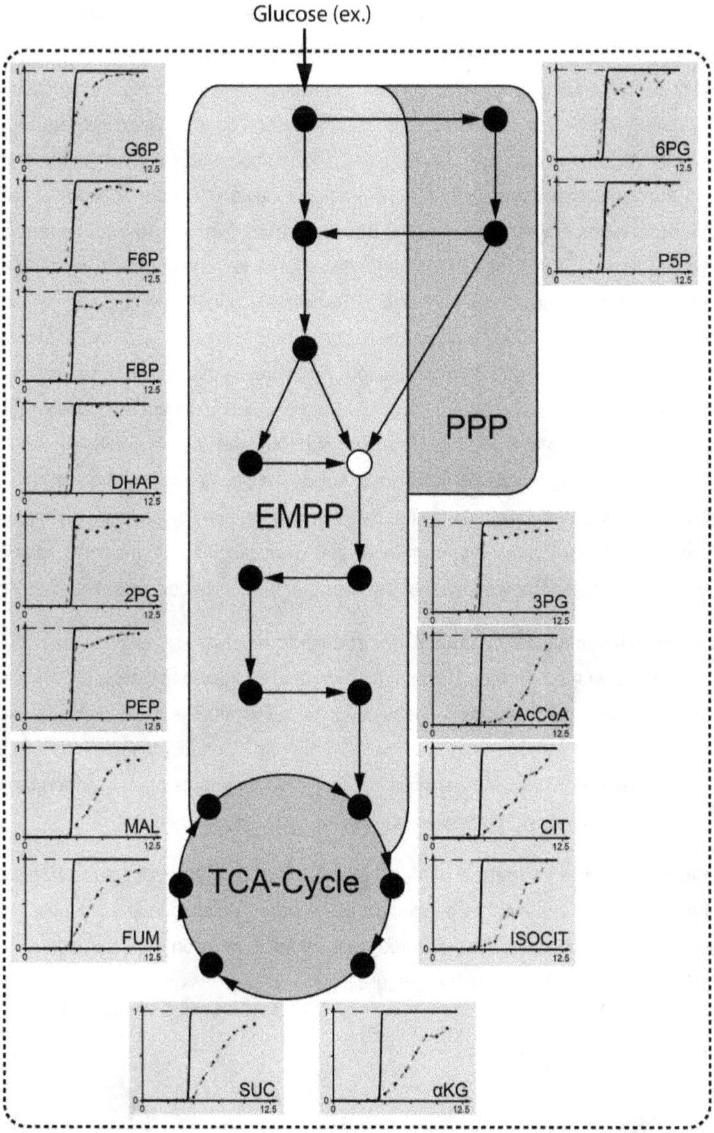

Figure 5.24 Comparison of measured and computed labeling patterns of selected metabolites measured for *E. coli* K12 DSM 2670 grown in fed-batch mode using [$^{13}C_6$] glucose as carbon source during the feeding phase. Three different patterns could be observed. Simulated and measured labeling states were almost alike for metabolites of glycolysis and PP pathway (highlighted in red). Labeling fractions of metabolites of the TCA cycle increased immediately, but the increase was slower and not completed at the end of the experiment (highlighted in green). The labeled fraction of AcCoA increased delayed if compared to the computed labeling pattern and was also not completed at the end of the experiment (highlighted in blue).

Summarizing, it seemed that *E. coli* utilized available carbon storages to maintain its energy level above 0.7. These results correlate with the findings of Reeve et al. (1984), who showed that cells of *E. coli* and *Salmonella thypimurium* degrade protein to prolong their survival during carbon starvation. However, as soon as all available carbon sources like glycogen, redundant proteins and extracellular organic acids have been consumed, the population might divide into oscillating subpopulations, thus exhibiting a drastically lowered AEC, as reported for yeast suspension and *Klebsiella aerogenes* (Hess & Boiteux 1968; Harrison 1976). The present data also indicated that a low AEC is not mandatorily a sign of cell death or errors during sample processing. Thus, cells can seemingly exhibit an extremely low AEC while being perfectly viable at the same time. This finding could be explained by subpopulations which oscillate between states of ATP utilization and regeneration. Additionally, the long-term fed-batch experiments indicated that an AEC below 0.3 equals a dying population, as the AEC did not fall below this threshold despite the formation of oscillating subpopulations. This finding has to be kept in mind, when evaluating metabolic datasets of larger cell populations, in order to avoid false conclusions. Noack & Wiechert (2014) already addressed this major issue of current methods of metabolite quantification, as they remarked that even small subpopulations, of less than 10 %, would result in serious variations of the average metabolite concentrations.

5.3 Central Carbon Metabolism of *Y. pseudotuberculosis*

As previously shown the developed quantitation method allowed generation of highly reliable datasets for Gram-positive and Gram-negative microorganisms. The method was now applied to compare the central carbon metabolism of the pathogen *Y. pseudotuberculosis* and the genetically close relative *E. coli*. Furthermore, four knock-out mutants of *Y. pseudotuberculosis* (*ΔarcA, ΔpdhR, ΔptsN, ΔpykF*), affected on their virulence, were analyzed.

5.3.1 Generation and Validation of Metabolic Profiles of *Y. pseudotuberculosis*

Y. pseudotuberculosis and the four deletion mutants were cultivated in shake flasks (n=3) using a minimal glucose medium. The cultures were sampled as soon as a biomass concentration of about 0.7 g/L had been reached (Figure 5.25).

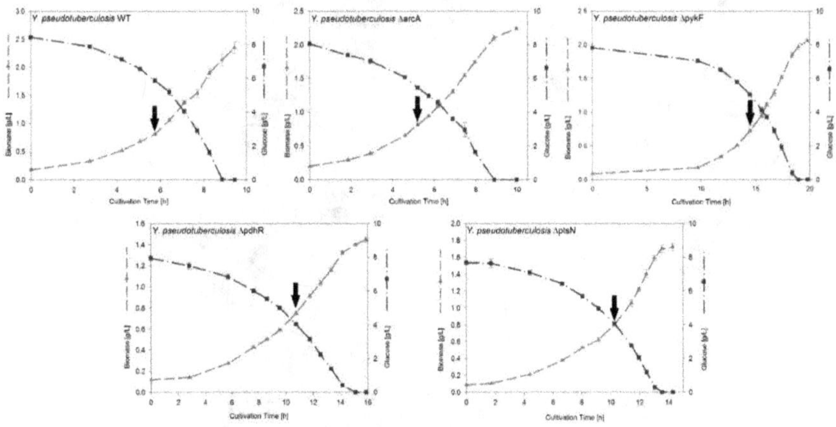

Figure 5.25: Cultivation profiles of *Y. pseudotuberculosis* wild type and the deletion strains *ΔarcA*, *ΔpykF*, *ΔpdhR* and *ΔptsN*. All strains were grown in baffled shake flasks (n=3). *Y. pseudotuberculosis* was cultivated under optimal conditions using defined medium with glucose as carbon source. Metabolite samples were taken during the late exponential phase as soon as a biomass of about 0.7 g/L had been reached (black arrows).

All strains exhibited exponential growth throughout the experiment. In addition, excess glucose was still present in the media at the time point of sampling. Samples were quenched and extracted as described above. Subsequently LC-MS/MS analysis was performed. The obtained metabolic datasets were validated by rigorous estimation of the reaction driving forces and Gibbs free energies (Figure 5.26). Neither the ED pathway nor the glyoxylate shunt seemed to be active in the wild type and its mutants, as KDPG and GA were not detected. These results are consistent with metabolic flux data (Bücker et al. 2014).

For almost all reactions, the obtained metabolite levels, driving forces (Figure 5.26) and Gibbs energies (Table 5.11) matched with the observed flux profiles (Bücker et al. 2014). As exception, the reaction from DHAP to GAP seemed to support the backward reaction of the triose phosphate isomerase and the phosphoglycerate mutase. It is likely that this single inconsistency is a consequence of an overestimated GAP pool. It is known that the GAP/DHAP ratio has to be lower than 0.048 to overcome the standard Gibbs energy of 7.5 kJ/mol (Cornish-Bowden 1981). The peak of GAP in the chromatogram, however, was small and broad, thus making a precise differentiation between background noise and signal difficult. Thus, further optimization of the method is needed to allow precise quantification of the GAP pool size, also other data sets (Luo et al. 2007; van der Werf et al. 2008) show this potential problem. Other, so far unrecognized sources of error remain also possible.

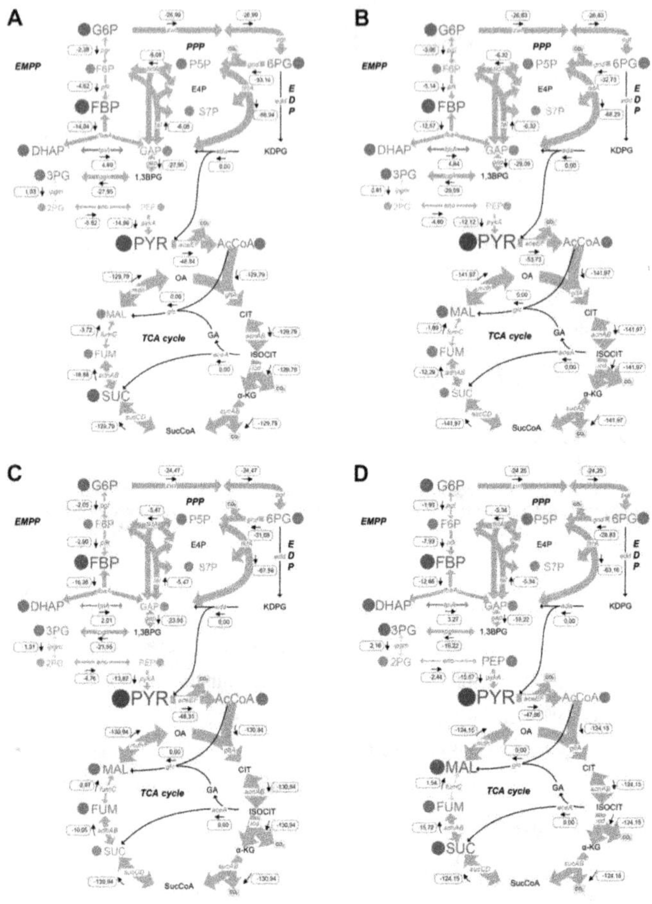

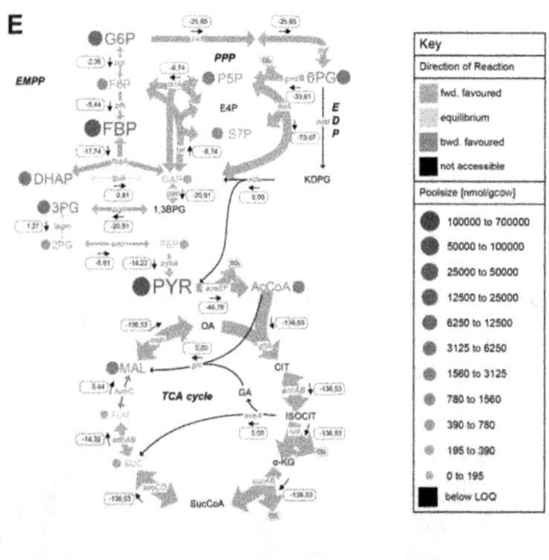

Figure 5.26: Evaluation of metabolic datasets of *Y. pseudotuberculosis ΔarcA* (A), *ΔpdhR* (B), *ΔptsN* (C), *ΔpykF* (D) and *Y. pseudotuberculosis* WT (E) by thermodynamic constraints and driving force, which were computed from the measured metabolite concentrations. Color and size of the letters of the metabolites and the adjacent circles represent the measured pool size. The color of the connecting arrows indicates in which direction the reaction would proceed for the measured metabolite concentrations, while the adjacent values represent the computed Gibbs energies for the corresponding reaction. Negative Gibbs energies support a forward directed reaction, positive energies result in a back reaction. The microorganisms were grown in baffled shake flasks containing minimal medium with glucose as the carbon source. Cultures were sampled during exponential growth at a biomass of about 0.7 g/L. GA and KDPG could not be found in the metabolic samples due to the inactivity of the ED pathway and the GA cycle. Consequently, no driving forces were calculated for reactions using these two metabolites.

Table 5.11: Comparison of the standard Gibbs energy (ΔG'') of reactions of the central carbon metabolism to Gibbs energies (ΔG) computed of the measured metabolite pools by application of Equation 3.2. Standard Gibbs energies were obtained from the database for thermodynamic properties (Li et al. 2011). Metabolite levels were obtained from samples of *E. coli* K12 DSM 2670, the wild type of *Y. pseudotuberculosis* and the *Y. pseudotuberculosis* knockout strains *ΔarcA*, *ΔpdhR*, *ΔptsN* and *ΔpykF*. All cultivations were performed in baffled shake flasks (n=3), using the corresponding minimal medium with glucose as the carbon source. Metabolite samples were taken during exponential growth phase, as soon as a biomass of 0.7 g/L had been reached.

Pathway	Reaction	ΔG *E. coli K12* [kJ/mol]	ΔG *Y. pseudo.* WT [kJ/mol]	ΔG *ΔarcA* [kJ/mol]	ΔG *ΔptsN* [kJ/mol]	ΔG *ΔpdhR* [kJ/mol]	ΔG *ΔpykF* [kJ/mol]	ΔG'' [kJ/mol]	Reference
EMP	G6P → F6P	-1.6	-2.4	-2.4	-2.1	-3.1	-1.9	2.78	Li et al., 2011
	F6P → FBP	-13.0	-5.4	-4.6	-2.9	-5.1	-7.9	-15.62	Li et al., 2011
	FBP → GAP + DHAP	-13.2	-17.7	-14.0	-16.4	-12.6	-12.7	24.64	Li et al., 2011
	DHAP → GAP	3.4	0.8	4.7	2.0	4.8	3.3	7.57	Li et al., 2011
	GAP → 1,3-bPG → 3PG	-15.5	-20.9	-28.0	-23.5	-29.1	-19.2	-19	Li et al., 2011
	3PG → 2PG	1.8	1.3	1.0	1.3	0.8	2.2	6.35	Li et al., 2011
	2PG → PEP	-1.7	-5.8	-5.8	-4.8	-4.6	-2.4	-4.47	Li et al., 2011
	PEP → PYR	-23.6	-14.2	-15.0	-13.8	-12.1	-15.6	-27.18	Li et al., 2011
	PYR → AcCoA	-34.5	-46.8	-48.8	-48.3	-53.7	-47.9	-39.26	Li et al., 2011
PP	G6P → 6PG	-27.5	-25.7	-27.0	-24.5	-26.8	-24.2	-21.89	Li et al., 2011
	6PG → P5P	-29.8	-33.6	-33.1	-31.1	-32.7	-28.8	1.23	Li et al., 2011
	2 6PG → GAP + S7P	-53.8	-73.1	-68.9	-67.6	-68.3	-63.2	-0.62	Li et al., 2011
	GAP + S7P → F6P + GAP	-6.4	-6.7	-6.1	-5.5	-6.3	-5.3	-6.08	Li et al., 2011
TCA cycle	MAL → SUC	-121.2	-136.5	-129.8	-130.9	-142.0	-124.2	-8.56	Li et al., 2011; Metacyc, 2015
	SUC → FUM	-12.9	-14.3	-18.9	-10.0	-12.3	-15.7	-0.59	Li et al., 2011
	FUM → MAL	0.4	3.4	-3.7	-3.0	-1.9	1.5	-3.52	Li et al., 2011

5.3.2 Comparison of the Core Metabolism of *Y. pseudotuberculosis* and *E. coli*

The metabolome of *Y. pseudotuberculosis* was further compared to *E. coli*, which had been analyzed during method development. In a first step, the ratios of the energy and redox equivalents were compared (Table 5.12).

Table 5.12: Adenylate phosphates pool (A(X)P), adenylate energy charge (AEC) and redox equivalent ratios measured for *E. coli* K12 DSM 2670, the *Y. pseudotuberculosis* wild type and the four mutant strains *Y. pseudotuberculosis ΔarcA*, *Y. pseudotuberculosis ΔpdhR*, *Y. pseudotuberculosis ΔptsN* and *Y. pseudotuberculosis ΔpykF*. All cultivations were performed in baffled shake flasks (n=3), using the corresponding minimal medium with glucose as the carbon source. Metabolite samples were taken during exponential growth phase, as soon as a biomass of 0.7 g/L had been reached.

	μ [h^{-1}]	A(X)P [μmol/g_{CDW}]	AEC [-]	NADH/NAD [-]	NADPH/NADP [-]
E. coli K12	0.13	10.52	0.81	0.26	1.06
Y. pseudotuberculosis	0.29	9.39	0.56	0.07	0.31
Y. pseudotuberculosis ΔarcA	0.28	9.53	0.46	0.04	0.27
Y. pseudotuberculosis ΔpykF	0.20	8.54	0.59	0.20	0.92
Y. pseudotuberculosis ΔpdhR	0.27	8.57	0.54	0.02	0.30
Y. pseudotuberculosis ΔptsN	0.23	10.21	0.46	0.16	0.64

Ratios of the redox equivalents resulted in values below one, thereby enhancing the driving forces of redox equivalent coupled reactions indicating an effective usage of catabolic pathways. Compared to the data obtained for *E. coli* the NADPH/NADP ratios of *Y. pseudotuberculosis* were substantially lower. The shortcoming of NADPH resulted in higher driving forces for NADPH producing reactions and might therefore result in an enhanced utilization of the oxidative part of the PP pathway. Additionally, two mutants (*ΔarcA*, *ΔpdhR*) exhibited an extremely low NADH/NAD ratio. This might be correlated with an upregulated respiratory chain, which was reported for *ΔarcA* and *ΔpdhR* mutants of *E. coli* (Spiro & Guest 1991; Trotter et al. 2011). In contrast to *E. coli*, *Y. pseudotuberculosis* and its knockout mutants exhibited an AEC below 0.6, while the overall adenylate phosphate concentration was about 9 μmol/g_{CDW}. This finding was unexpected, as the cells grew exponentially, while the AEC indicated rather starving cells. However, thermodynamic validation showed that the metabolite levels were physiologically meaningful. As cells were viable at the point of sampling, these results indicate that *Y. pseudotuberculosis* lowers its AEC, possibly to enhance the rate of glycolysis. It is known, that *Y. pseudotuberculosis* secretes large amounts of PYR into the surrounding medium (Bücker et al. 2014). This overflow might be the consequence of an enhanced flow through glycolysis, which could not be completely channeled into the TCA cycle. The high amounts of extracellular PYR might be a countermeasure against the reactive oxygen species, which are produced by macrophages to attack pathogens. It has been shown that PYR acts as a protectant against reactive oxygen species, such as hydrogen peroxide (Desagher et al. 1997; Troxell et al. 2014). Thus, *Y. pseudotuberculosis* might use the low AEC to stimulate the formation of PYR during enzootic cycles. To evaluate these hypotheses further, a closer look at the prominent differences between the wild types of *Y. pseudotuberculosis* and *E. coli* was taken and compared to the metabolite levels after deletion of *arcA*, *pdhR*, *ptsN* and *pykF*.

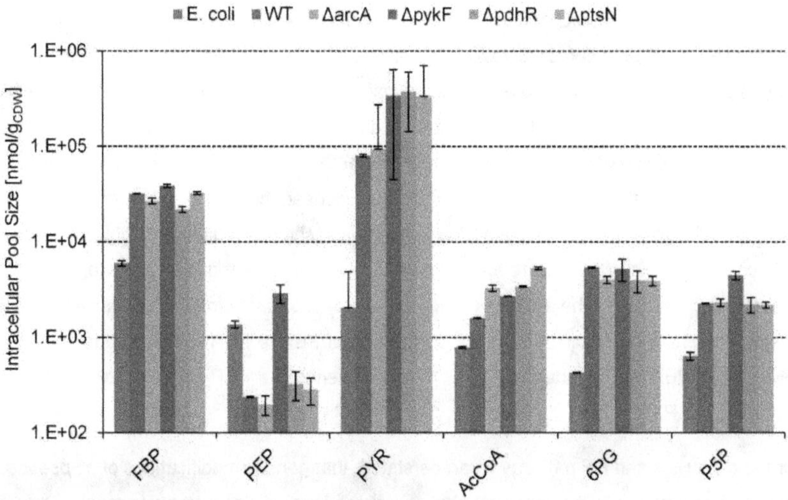

Figure 5.27: Metabolite pools of glycolysis and PP pathway (n=3), which differed greatly between *E. coli* K12 DSM 2670 and wild type *Y. pseudotuberculosis* or which were strongly affected by knockout of arcA, pdhR, ptsN and pykF in *Y. pseudotuberculosis.* All cultivations were performed in baffled shake flasks (n=3), using the corresponding minimal medium with glucose as the carbon source. Metabolite samples were taken during exponential growth phase, as soon as a biomass of 0.7 g/L had been reached.

Figure 5.27 depicts the pool sizes of the four mutants in comparison to the two wild types for relevant metabolites of glycolysis and the PP pathway. The majority of metabolite concentrations upstream of PYR remain at a constant level, almost equal to the pool sizes observed for *E. coli*. In contrast pools downstream of PYR, i.e. AcCoA (Figure 5.27) and the organic acids (Figure 5.28), were more strongly perturbed by the genetic alterations and differed widely from *E. coli*. Furthermore, the quantitative analysis identified 6 metabolites from glycolysis and PP pathway, which differed greatly between *E. coli* and *Y. pseudotuberculosis*, while the genetic alterations showed almost no effect. The intracellular FBP level in *Y. pseudotuberculosis* was five times higher than in *E. coli,* probably as a consequence of the low AEC of *Y. pseudotuberculosis*. The key enzyme of glycolysis, the phosphofructokinase, shows higher activity at low energy charges, as the enzyme is activated by AMP (Atkinson & Walton 1967). As consequence of the activation, a higher FBP level was measured. In addition, metabolite levels of 6PG and the pentose-5-phosphate pool were markedly higher compared to *E. coli*. This might be a direct consequence of a lowered NADPH/NADP ratio, found for *Y. pseudotuberculosis* (Table 5.12). The low NADPH/NAPD ratio favors reactions which produce NADPH, thus an elevated utilization of the oxidative PP pathway to synthesize NADPH seems likely. Consequently, the glucose-6-phopshate dehydrogenase and 6-phosphogluconate dehydrogenase rates were elevated, resulting in the elevated pools of 6PG and P5P. These findings were in good agreement with results of metabolic flux analysis, which showed an increased use of the PP pathway for *Y. pseudotuberculosis* (Bücker et al. 2014). However, the most prominent difference between *E. coli* and *Y. pseudotuberculosis* was the amount of intracellular PYR, which was found to be more than 40 times higher in all samples of *Y. pseudotuberculosis*. These large amounts of PYR even affected the adjacent AcCoA pool, which was also markedly elevated compared to the pool size found in *E. coli* as a consequence of the changed mass action ratio. The PEP pool, which precedes PYR in glycolysis, showed a different picture. Compared to *E. coli*, the levels of PEP were found to be substantially lowered in *Y. pseudotuberculosis* (Figure 5.27). This might be a consequence of the low abundance of intracellular ATP, as the lack of ATP shifts the equilibrium of the conversion of PEP to PYR to the product side. As a result, more PEP is converted to PYR, thereby lowering the PEP pool and simultaneously increasing the PYR pool.

For the glycolysis and PP pathway it can be stated, that genetic modifications of *Y. pseudotuberculosis* resulted only in minor perturbations of the intracellular metabolite levels. Only the deletion of pyruvate kinase led to a relevant change in the level of intracellular PEP. As a consequence of the deletion of pyruvate kinase F, the interconversion of PEP to PYR is slowed down, which probably caused a 10-fold increase of the intracellular PEP concentration. PEP itself is known to act as an inhibitor of glycolytic enzymes, which might explain the observed increase of metabolite pools of the preceding reactions of glycolysis found for the *ΔpykF* strain (Kelly & Turner 1969). *E. coli* reacts to the knockout of the pyruvate kinase by redirecting the flux through PEP carboxylase and the malic enzyme, thus circumventing the arising bottleneck (Al Zaid Siddiquee et al. 2004). The closely related *Y. pseudotuberculosis* seems to react in a similar way. This hypothesis is supported by two observations. Firstly, even though the pyruvate secretion was lowered by knockout of pyruvate kinase, the intracellular concentration of pyruvate was three times higher in the mutant compared to the wild type. Secondly, a strongly increased malate pool (20-fold) (Figure 5.28), which is likely a consequence of a redirected flux of PEP to PYR via PEP carboxylase and malic enzyme. The 20-fold higher intracellular malate concentration additionally resulted in substantially higher intracellular concentrations of SUC and FUM as a consequence of backlog in the TCA cycle. Thus, the metabolite measurements indicate, that the deletion of pyruvate kinase F had no negative effect on the PYR formation of *Y. pseudotuberculosis*. It is likely that the missing pyruvate kinase F was compensated by the utilization of PEP carboxylase and malic enzyme.

In response to the deletion *pykF*, metabolite pools of the TCA cycle and PYR itself showed larger differences. The intracellular PYR concentration of the strains *ΔpykF, ΔpdhR,* and *ΔptsN* was three times higher than in the wild type of *Y. pseudotuberculosis*, while the PYR level of the *ΔarcA* strain remained constant (Figure 5.27). Previously, it has been reported for *E. coli* that the phosphorylation state of $EIIA^{Glc}$ is coupled to the growth rate and is furthermore strongly dependent on the PEP/PYR-ratio (Hogema et al. 1998; Gabor et al. 2011). As the PEP to pyruvate ratios of *Y. pseudotuberculosis* were apparently perturbed by the genetic mutations, the influence of the altered ratios on the growth rate had to be evaluated. The PEP/PYR-ratios calculated for the knockout mutants can be seen in Table 5.13. On the one hand, the *ΔpykF* strain exhibits the highest ratio ($8.55 \cdot 10^{-3}$), as a consequence of its drastically increased PEP pool. On the other hand, the knockout of the metabolic regulators ArcA, PtsN and PdhR resulted in substantially lowered ratios. Bettenbrock et al. (2007) showed that a PEP/PYR ratio below 0.4 is correlated with low phosphorylation states of $EIIA^{Glc}$ and high growth rates in *E. coli*. The PEP/PYR ratios of the knockout strains of *Y. pseudotuberculosis* did not exceed this critical value (Table 5.13), thus a negative influence of the altered PEP and PYR pools of *Y. pseudotuberculosis* on growth rate and virulence seems unlikely.

Table 5.13: PEP/PYR ratios derived from measured intracellular pools of *Y. pseudotuberculosis* wild type and *Y. pseudotuberculosis* mutant strains *ΔarcA, ΔpdhR, ΔptsN* and *ΔpykF*. Cultivations were performed in baffled shake flasks (n=3), using the corresponding minimal medium with glucose as the carbon source. Metabolite samples were taken during exponential growth phase, as soon as a biomass of 0.7 g/L had been reached.

	WT	*ΔarcA*	*ΔptsN*	*ΔpdhR*	*ΔpykF*
PEP/PYR [-]	0.0030	0.0021	0.0009	0.0009	0.0085
μ [h^{-1}]	0.29	0.28	0.23	0.27	0.20

The levels of TCA cycle intermediates showed a different picture (Figure 5.28). Here, strong fluctuations in pool size resulted from genetic modifications of *Y. pseudotuberculosis*. The intracellular concentrations of CIT, ISOCIT and αKG, however, were so low that the pool size could not be precisely estimated.

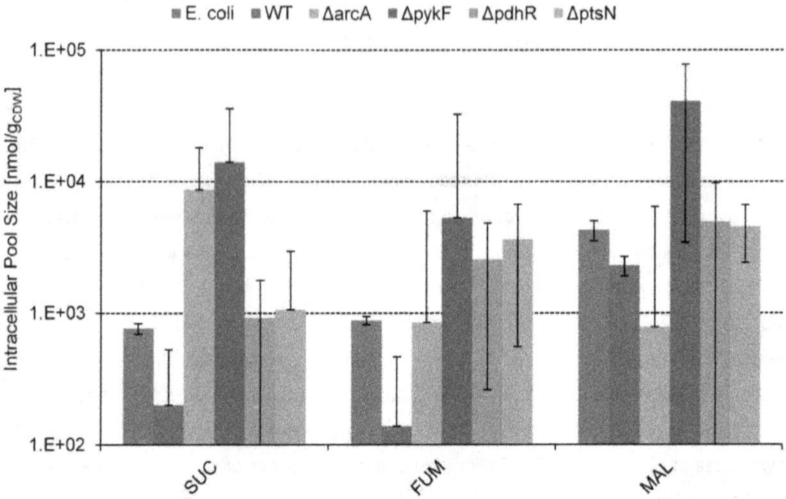

Figure 5.28: Metabolite pools of the TCA cycle (n=3), which differ greatly between *E. coli* K12 DSM 2670 and *Y. pseudotuberculosis* or which were strongly affected by knockout of arcA, pdhR, ptsN and pykF in *Y. pseudotuberculosis*. Cultivations were performed in baffled shake flasks (n=3), using the corresponding minimal medium with glucose as the carbon source. Metabolite samples were taken during exponential growth phase, as soon as a biomass of 0.7 g/L had been reached.

The intracellular metabolite levels showed that the pools of SUC, FUM and MAL were lower in wild type *Y. pseudotuberculosis* than in wild type *E. coli*. These findings correlate with the results of metabolic flux analysis, which indicate a higher TCA flux for *E. coli* (Hiller 2006; Bücker et al. 2014). Quantitative analysis of the metabolite levels of SUC, FUM and MAL showed that these pools were several times higher in the knockout mutants. Thereby the metabolic profile resembled more the metabolite levels of *E. coli* than that of the wild type of *Y. pseudotuberculosis*. The increase of these pools as a consequence of the genetic modifications, however, seemed not the result of an elevated flux through the TCA cycle. Thus, a closer look at the genetic modifications had to be taken.

The redox regulator ArcA (anoxic redox control) is known to control TCA cycle fluxes in *E. coli* during aerobic batch growth on glucose (Perrenoud & Sauer 2005). Apart from modulating the transcription of enzymes which are actively participating in the TCA cycle, ArcA is responsible for the fine tuning of the respiratory chain. The regulator is activated under microaerobic conditions and then represses the transcription of cytochrome oxidase while simultaneously activating transcription of cytochrome d oxidase, which shows a higher affinity to oxygen (Spiro & Guest 1991). Furthermore it has been shown that ArcA is responsible for the flux distribution ratio in *E. coli* at the branching point between acetate secretion and TCA cycle (Haverkorn van

Rijsewijk et al. 2011). Response of *Y. pseudotuberculosis* to the deletion of *arcA* varied slightly from the response of *E. coli*. Redox equivalents ratios of the *ΔarcA* strain (Table 5.12) were slightly lowered if compared to redox equivalent ratios of wild type *Y. pseudotuberculosis*. This finding indicates, in good agreement with earlier findings (Spiro & Guest 1991; Haverkorn van Rijsewijk et al. 2011), an upregulated respiratory chain. Additionally, the accumulation of intracellular SUC and FUM, combined with a lowered MAL pool (Figure 5.28), could be a consequence of higher transcription rates of the enzymes 2-oxoglutarate-, succinate-dehydrogenase and fumarase due to the lacking repression by activated ArcA. In contrast to the results of Haverkorn van Rijsewijk et al. (2011), which were obtained for *ΔarcA* mutants of *E. coli*, no indication for an upregulation of other enzymes of the TCA cycle were found. A possible explanation could be that the *Y. pseudotuberculosis* wild type already operates at a high efficiency for the reactions leading from MAL to αKG, thus an upregulation of the partaking enzymes had no effect on the metabolite levels. This result could also explain why the intracellular concentrations of CIT, ISOCIT and αKG could not be determined by application of the differential method, as a high enzymatic activity of the enzymes would inevitably keep the concentrations of CIT, ISOCIT and αKG close to zero.

Deletion of the pyruvate kinase F (PykF), one of the core enzymes of the pyruvate node, resulted in drastic changes in the pool sizes of the central carbon metabolism, as discussed above. It became apparent that the knockout of *pykF* led to a backlog of metabolites of the preceding reactions in EMP- and PP pathway. The ratios of substrate to product, however, remained constant. Consequently, most conversion rates of glycolysis and PP pathway were found to be unchanged. The conversion of PEP to PYR by the utilization of phosphoenolpyruvate carboxylase and malic enzyme resulted in a 20-fold increased MAL pool. The increase of the MAL level resulted in a further backlog, leading to increasing pools of the preceding metabolites of the TCA cycle, namely SUC and FUM (Figure 5.28). The co-infection experiments showed that the deletion of *pykF* was a disadvantage for *Y. pseudotuberculosis* in the deeper tissues, while it was of no effect during first colonization of the host organism (Bücker et al. 2014). The cause of these results could be the continuously high demand of PYR which forces the cells to circumvent the pyruvate kinase by utilization of PEP-carboxylase and malic enzyme. The usage of this futile cycle results in the loss of 1 mol ATP per generated mol of PYR, a waste of energy that consequently reduces the virulence of the knockout strain in deeper tissues. Energy efficiency is crucial in later phases of infection, while futile cycles can be compensated under glucose excess during colonization of the host. This conclusion is also supported by the relatively high AEC of the PykF knockout strain (Table 5.12).

The pyruvate dehydrogenase complex regulator (PdhR) functions as a transcriptional regulator for the pyruvate dehydrogenase complex (PDHC), which catalyzes the conversion from PYR to AcCoA. Recently it has been shown that PdhR additionally acts as a sensor of the intracellular pyruvate concentration and that it fulfills a key role in the reprogramming of *E. coli* during the transition from an anaerobic to a microaerobic environment (Trotter et al. 2011). Furthermore, it has been shown that, apart from controlling the PDHC, PdhR has additional functionality as a transcriptional regulator of enzymes of the TCA cycle, the glyoxylate bypass, the respiratory chain and even links cell replication to the nutritional status of the cell (Göhler et al. 2011). A closer look at the changes in metabolite pool sizes reveals that the central carbon metabolism of the *ΔpdhR* strain was successfully perturbed at the pyruvate node, as the concentrations of PYR, AcCoA and the metabolites of the TCA cycle differ greatly from the concentrations of the wild type (Figure 5.27, Figure 5.28). In good agreement with the reported upregulation of the NADH dehydrogenase and the respiratory chain (Trotter et al. 2011), the NADH/NAD ratio of the *ΔpdhR* mutant drops to a value of 0.02 (Table 5.12). Moreover, increased metabolite levels of SUC, FUM and MAL of the PdhR knockout strain could be a consequence of elevated conversion rates of the TCA cycle. Göhler et al. (2011) reported higher transcription rates for most enzymes of the TCA for *E. coli* cultures grown on pyruvate, while transcription of succinate dehydrogenase was lowered at the same time, thus supporting the results of the metabolome analysis performed in this work.

Previous transcriptome and flux analyses of *Y. pseudotuberculosis* had shown that deletions of the transcriptional and post-transcriptional regulators RovA, CsrA and Crp strongly affect the central carbon metabolism (Bücker et al. 2014). Especially pyruvate metabolism and the entrance of carbon into the TCA cycle were perturbed by these mutations, thus leading to a reduced virulence of the pathogen. In a next step, knockouts of the central enzyme of the pyruvate-TCA cycle node, pyruvate kinase (*ΔpykF*), and essential regulators of this metabolic branching point *(ΔarcA, ΔpdhR, ΔptsN)* were therefore established and analyzed in respect of the resulting changes in the central metabolism and virulence. Overall, it can be noted that the metabolome analysis indicates a perturbation of pyruvate metabolism and TCA cycle of *Y. pseudotuberculosis* as consequence of the genetic modifications. Furthermore, it was shown, that of the four introduced mutations only *ΔpykF* exhibited changes throughout the complete set of metabolites, whereas alterations in *ΔarcA*, *ΔptsN* and *ΔpdhR* were restricted to the TCA cycle metabolites, PYR and AcCoA. The PEP/PYR-ratio, which was previously identified as a possible target to reduce the virulence of *Y. pseudotuberculosis*, showed no negative influence on growth rate or virulence. However, it could be shown that the *ΔpykF* strain utilized a futile cycle to maintain a high PYR pool. This loss of energy results in a disadvantage for the knockout strain when competing for resources. The deletions of *ptsN* and *pdhR* resulted only in minor changes of metabolome and driving forces, therefore a link between virulence and metabolism could not be established.

However, it has to be noted that the metabolome of the wild type and all four knockout mutants was sampled under glucose excess conditions and from fully aerobic cultures. These conditions are certainly not representative for conditions in the lymphatic system, the liver or the spleen. Recently, it was demonstrated that the ability to convert SUC via FUM, MAL and OA to PYR plays a major role in the virulence of *Salmonella enterica* (Yimga et al. 2006; Mercado-Lubo et al. 2008; Mercado-Lubo et al. 2009). It was reported that the complete loss of one of these conversion steps rendered *S. enterica* growing in all deeper tissues, but at the same time being fully avirulent. To explain these findings, it was hypothesized that *S. enterica* metabolizes succinate, arginine or ornithine, which are still available in deeper tissues and phagocytes, as soon as glucose becomes scarce. It was proposed, that maintenance of high PYR levels is important for the virulence of *S. enterica*, as it is a precursor for amino acids and AcCoA. The metabolite levels of SUC, FUM, MAL and PYR, which were measured for *Y. pseudotuberculosis* and its knockout mutants show large variations, thereby indicating a deregulating effect of the gene deletions. Hence, it is likely that metabolite quantification and flux analysis of *Y. pseudotuberculosis* strains, grown using succinate, arginine or ornithine as the sole carbon source, would give a more representative picture of the metabolic changes induced by the gene deletions. The additional information might enable elucidation of the reduced virulence of the *ΔarcA*, *ΔptsN* and *ΔpykF* strain in lymph nodes, liver and spleen.

5.4 Stress Induced Changes in the Metabolome of *B. megaterium*

The following chapter elucidates the response of the central carbon metabolome of *B. megaterium* to stress, by temperature or by salt concentration. For this investigation, metabolite analysis was performed using cultures growing at temperatures ranging from 15 °C to 45 °C as well as cultures growing in media containing 0.6 M and 1.2 M of sodium chloride. The obtained metabolic datasets were compared to the metabolite levels of *B. megaterium*, grown under optimal conditions (37 °C, 0 M), to elucidate the effects of stress on metabolism.

5.4.1 Generation and Validation of Metabolic Profiles of *B. megaterium*

B. megaterium was grown in shake flasks as described above. All cultivations were performed as triplicate and sampled as soon as a biomass level of 1.0 g/L had been reached. Biomass and glucose concentration were monitored throughout the complete cultivation (Figure 5.29).

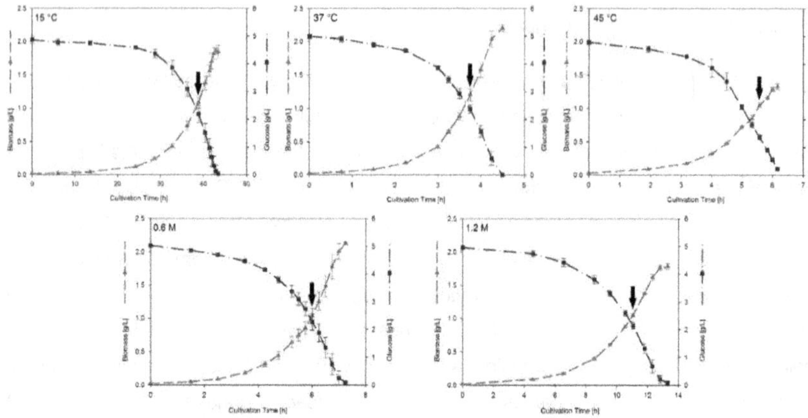

Figure 5.29: Cultivation profiles of *B. megaterium* grown in shake flasks (n=3). *B. megaterium* was cultivated under optimal conditions using modified M9 medium. To stress the bacterium, different salt concentrations (0.6 M, 1.2 M) were added to the medium or the temperature was changed into suboptimal ranges (15 °C, 45 °C). Metabolite samples were taken during exponential growth, as soon as a biomass of 1 g/L had been reached (black arrows).

Obviously, *B. megaterium* exhibited exponential growth under all conditions. However, it can be seen that the different growth conditions resulted in different growth rates. Furthermore, it can be seen that the glucose concentration was above 2.0 g/L at the time of metabolome sampling. In addition, the samples were taken at low biomass concentrations to assure sufficient oxygen saturation of the culture broth throughout the complete cultivation. Thereby, it should be assured that measured fluctuations in the metabolic datasets were exclusively results of the intended variations of temperature or salt concentration. The samples were then extracted and analyzed as previously described. In accordance with results of the metabolic flux analysis (Godard 2015), KDPG and GA could not be detected. Thus, ED pathway and GA shunt seemed to be inactive in *B. megaterium* under all conditions. As can be seen in Figure 5.30 and Table 5.14, validation of the measured metabolite levels by driving force and Gibbs energies indicated that the central carbon metabolism of *B. megaterium* was functional despite the stressing conditions.

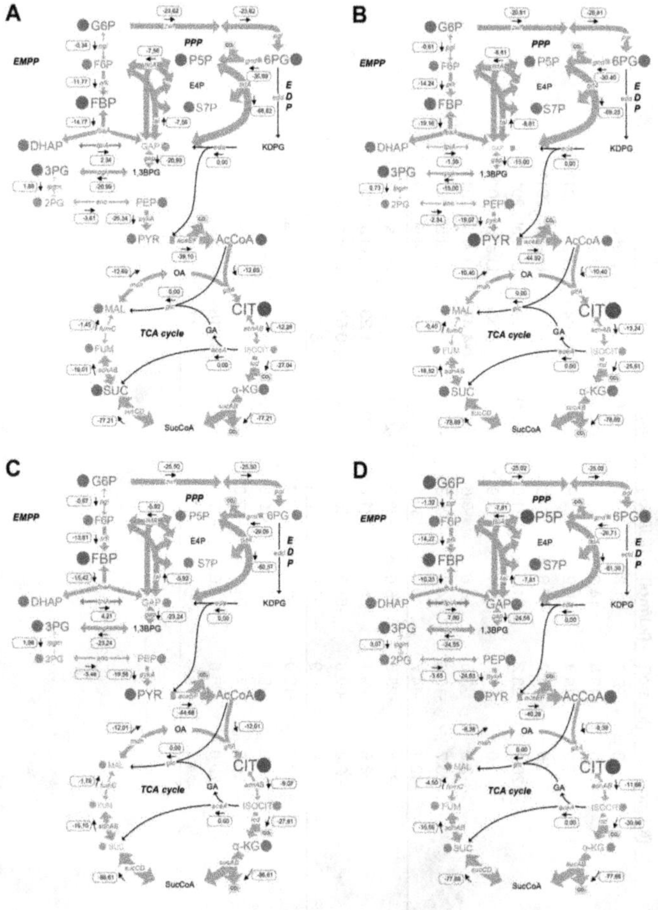

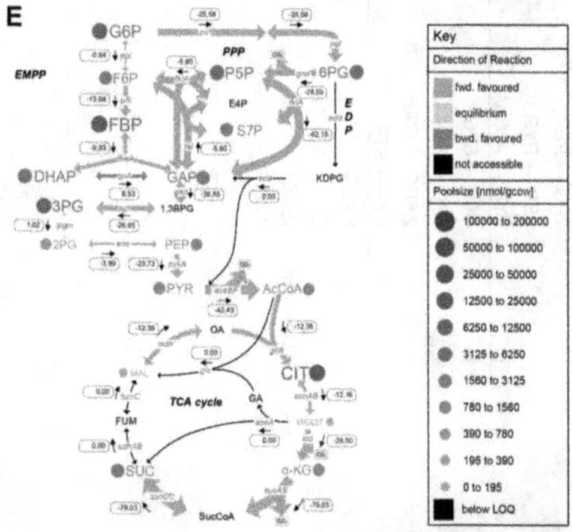

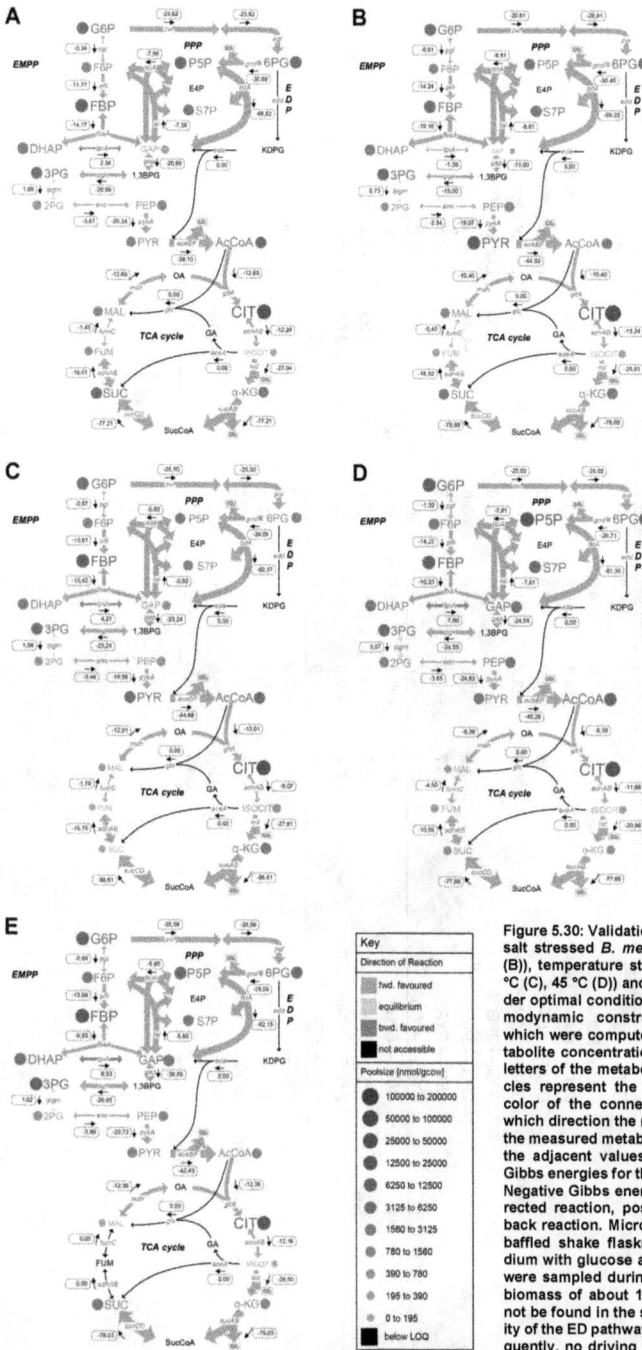

Figure 5.30: Validation of metabolic datasets of salt stressed *B. megaterium* (0.6 M (A), 1.2 M (B)), temperature stressed *B. megaterium* (15 °C (C), 45 °C (D)) and *B. megaterium* grown under optimal conditions (0 M, 37 °C (E)) by thermodynamic constraints and driving force, which were computed from the measured metabolite concentrations. Color and size of the letters of the metabolites and the adjacent circles represent the measured pool size. The color of the connecting arrows indicates in which direction the reaction would proceed for the measured metabolite concentrations, while the adjacent values represent the computed Gibbs energies for the corresponding reaction. Negative Gibbs energies support a forward directed reaction, positive energies result in a back reaction. Microorganisms were grown in baffled shake flasks containing minimal medium with glucose as carbon source. Cultures were sampled during exponential growth at a biomass of about 1 g/L. GA and KDPG could not be found in the samples due to the inactivity of the ED pathway and the GA cycle. Consequently, no driving forces were calculated for reactions using these two metabolites.

Table 5.14: Comparison of the standard Gibbs energy ($\Delta G''$) of reactions of the central carbon metabolism to Gibbs energies (ΔG) computed of the measured metabolite pools by application of Equation 3.2. Standard Gibbs energies were obtained from the database for thermodynamic properties (Li et al. 2011). Metabolite levels were obtained from samples of *B. megaterium* grown under optimal conditions (37 °C, 0 M NaCl) and under temperature induced (15 °C, 45 °C) or salt induced (0.6 M, 1.2 M) stress. All cultivations were performed in baffled shake flasks (n=3), using minimal medium with glucose as the carbon source. Metabolite samples were taken during exponential growth phase at a biomass of 1 g/L.

Pathway	Reaction	ΔG *B. megaterium* [kJ/mol]	ΔG 15 °C [kJ/mol]	ΔG 45 °C [kJ/mol]	ΔG 0.6 M [kJ/mol]	ΔG 1.2 M [kJ/mol]	ΔG'' [kJ/mol]	Reference
EMP	G6P → F6P	-0.8	-0.7	-1.3	-0.3	-0.6	2.78	Li et al., 2011
	F6P → FBP	-13.0	-13.8	-14.3	-11.8	-14.2	-15.62	Li et al., 2011
	FBP → GAP + DHAP	-9.9	-15.4	-10.2	-14.8	-19.2	24.64	Li et al., 2011
	DHAP → GAP	6.5	4.2	7.8	2.3	-1.6	7.57	Li et al., 2011
	GAP → 1,3-bPG → 3PG	-26.9	-23.2	-24.5	-21.0	-15.0	-19	Li et al., 2011
	3PG → 2PG	1.0	1.1	3.1	1.9	0.7	6.35	Li et al., 2011
	2PG → PEP	-3.9	-3.5	-3.6	-3.6	-2.3	-4.47	Li et al., 2011
	PEP → PYR	-23.7	-19.6	-24.6	-25.3	-19.1	-27.18	Li et al., 2011
	PYR → AcCoA	-42.4	-44.7	-40.3	-39.1	-44.9	-39.26	Li et al., 2011
PP	G6P → 6PG	-25.6	-25.5	-25.0	-23.6	-20.8	-21.89	Li et al., 2011
	6PG → P5P	-28.6	-29.1	-26.7	-31.0	-30.4	1.23	Li et al., 2011
	2 6PG → GAP + S7P	-62.2	-60.6	-61.3	-68.8	-69.2	-0.62	Li et al., 2011
	GAP + S7P → F6P + GAP	-5.8	-5.9	-7.8	-7.6	-8.8	-6.08	Li et al., 2011
TCA cycle	MAL → CIT	-12.4	-12.0	-8.4	-12.7	-10.4	-8.56	Li et al., 2011
	CIT → ISOCIT	-12.2	-9.1	-11.7	-12.3	-13.2	0.92	Metacyc, 2015
	ISOCIT → AKG	-26.5	-27.6	-31.0	-27.0	-25.6	-3.29	Li et al., 2011
	AKG → SUC	-79.0	-86.6	-77.9	-77.2	-78.9	-43.18	Metacyc, 2015
	SUC → MAL	-24.4	-16.9	-20.1	-20.5	-19.0	-4.11	Li et al., 2011

Similar to the validation of *Y. pseudotuberculosis* and *E. coli*, the ratio of GAP and DHAP was the only reaction which did not match with the expected flux direction. As described above, this could be a consequence of an overestimation of the GAP pool. Subsequently, the energetic state and the redox state of the cells were analyzed (Table 5.15).

Table 5.15: Adenylate phosphate pools, adenylate energy charge and redox equivalent ratios measured for *B. megaterium* grown under optimal conditions (37 °C, 0M NaCl) and under temperature induced (15 °C, 45 °C) or salt induced (0.6 M, 1.2 M) stress. All cultivations were performed in baffled shake flasks (n=3), using minimal medium with glucose as the carbon source. Metabolite samples were taken during exponential growth phase at a biomass of 1 g/L.

	μ [h^{-1}]	A(X)P [μmol/g_{CDW}]	AEC [-]	NADH/NAD [-]	NADPH/NADP [-]
B. megaterium at 37 °C, 0 M	1.19	12.00	0.84	0.01	0.52
B. megaterium at 15 °C, 0 M	0.15	6.03	0.88	0.01	0.81
B. megaterium at 45 °C, 0 M	0.65	14.67	0.84	0.02	0.57
B. megaterium at 37 °C, 0.6 M	0.69	11.91	0.77	0.03	0.43
B. megaterium at 37 °C,1.2 M	0.39	12.26	0.86	0.02	1.06

AEC values were never below 0.75, thus all cells were viable at the point of sampling despite the imposed stress (Table 5.15). These values were found to be in good agreement with data previously published for *B. megaterium* by Chapman et al. (1971) and Rubia et al. (1986). The overall concentrations of adenylate phosphates per cellular dry weight, however, correlated with the applied cultivation temperature. These results indicate that *B. megaterium* adapts the available adenylate phosphate concentration to fit the actual needs of the cell. Low temperatures result in lowered reaction velocities (Beales 2004). Therefore, high amounts of adenylate phosphates become redundant, especially since a high amount of AMP would lead to a lowered AEC and thus in a downregulation of anabolic reactions. Hence, the results indicate that the active regulation of the intracellular adenylate phosphate pool is not limited to starvation processes of *E. coli* (Chapman et al. 1971), but is a reaction of all microorganisms to environmental changes. In addition, the NADH/NAD ratio of *B. megaterium* remained stable at values below 0.05. The NADH/NAD measured ratios were in good agreement with data published for *B. megaterium* (Setlow & Setlow 1977). A low NADH/NAD ratio supports effective and rapid metabolization of glucose via glycolysis and TCA cycle, as it controls important parameters, such as the activity of the PDH and the respiratory chain (Pettit et al. 1975; Vemuri et al. 2007). Concluding, it can be stated that the metabolic datasets were successfully validated. Additionally, the AEC-values as well as the ratios of redox-equivalents lie well within the range previously published by Chapman et al. (1971) and Bolten (2010). Thus, the data were now applied to further investigate the influence of temperature and salt stress on the core metabolism of *B. megaterium*. However, as temperature shifts result in changes of the reaction rate and protein degradation, while changes in the salt concentration lead to problems with osmolarity, both conditions provoke different cellular responses.

5.4.2 Response of *B. megaterium* to Temperature Induced Stress

A first impact of the differing cultivation temperatures could be seen in the specific growth rates, which were drastically lowered compared to the specific growth rate under optimal conditions (37 °C). The PEP/PYR ratio at 15 °C and 45 °C was 0.17 and 0.92, respectively (Table 5.16). Consequently, a negative influence of the PEP/PYR-ratio on the phosphorylation state of the Glc-PTS could be excluded for cultivation at 15 °C. However, the glucose uptake of *B. megaterium* was decreased at 15 °C, despite the feasible PEP/PYR ratio. This might be a consequence of reduced affinity of membrane transporters to their substrates and stiffening of the cellular membrane at low temperatures (Nedwell 1999; Los & Murata 2004). The high ratio of 0.92 at 45 °C, on the other hand, indicated a lowered phosphorylation state of the Glc-PTS. The resulting decrease in glucose uptake is possibly one of the reasons for the observed decrease of the growth rate at higher temperatures.

Table 5.16: Growth rates and measured PEP/PYR ratios exhibited by *B. megaterium* grown at optimal conditions (37 °C) and at stress inducing temperatures (15 °C, 45 °C). Cultivations were performed in baffled shake flasks (n=3), using minimal medium with glucose as carbon source. Metabolite samples were taken during exponential growth phase at a biomass of 1 g/L.

	37 °C	15 °C	45 °C
PEP/PYR [-]	0.62	0.17	0.92
μ [h^{-1}]	1.19	0.15	0.65

To identify further effects of extreme temperatures on the central carbon metabolism of *B. megaterium*, a closer look at the complete metabolic profile was taken (Figure 5.31). The majority of the measured metabolites of PP and EMP pathway increased with increasing temperature. This might be the result of a change in reaction velocity. The reaction velocity of biochemical reaction depends on the temperature of the surrounding medium. An increase in temperature of 10 K will result in a two to three-fold increase of the reaction velocity. On the other hand, high temperatures lead to reduction or even complete loss of enzymatic activity due to protein degradation, thereby reducing the velocity of intracellular glucose conversion. Both phenomena have to be considered, when analyzing the development of the intracellular metabolite pools as a reaction towards the cultivation of *B. megaterium* at 15 °C and 45 °C, respectively. The analysis is additionally aggravated by differing heat stabilities of the partaking enzymes, as it is hard to distinguish if the metabolite pool is changing as a consequence of altered enzyme activities of incoming or outgoing reactions. However, as most metabolite concentrations increased with increasing temperatures, the effect of protein denaturation seemed to be negligible.

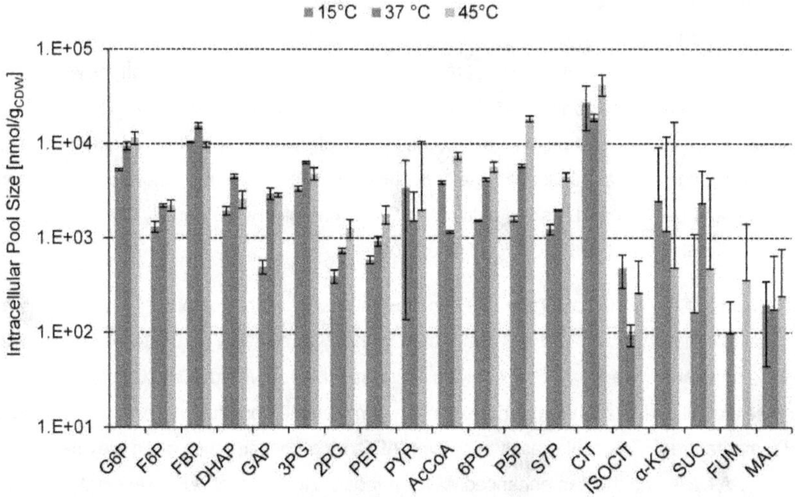

Figure 5.31: Effect of different temperatures (15°C, 37 °C, 45 °C) on metabolite pools of the central carbon metabolism measured in *B. megaterium*. *B. megaterium* was grown in shake flasks (n=3) containing modified M9 medium with glucose as carbon source. Metabolite samples were taken during exponential growth phase at a biomass of 1 g/L.

In contrast to the majority of metabolites, the pools of GAP and PYR respond differently to the change in temperature. Both pools additionally influence the adjacent metabolite levels, as incoming and outgoing reactions respond to the change in metabolite concentrations. Thus, the measured intracellular metabolite concentrations, driving force and flux analysis were combined to identify the cause of the variable PYR and GAP levels.

Looking at Gibbs energies (Table 5.14), it became obvious that *B. megaterium* successfully compensated most effects of temperature induced changes of the carbon metabolism. It can be seen that the cleavage of FBP to DHAP and GAP at a temperature of 15 °C is the only reaction of glycolysis and pentose phosphate pathway which is affected by changing metabolite levels. These results are in good agreement with the relative flux maps computed for growth under temperature induced stress. Flux analysis showed that the synthesis of glycogen and peptidoglycan is upregulated at low temperatures, thereby lowering the flux into EMP- and PP pathway. As a result of the reduced flux, *B. megaterium* lowers the rates of nucleotide and amino acid synthesis from intermediates of the PP pathway, thus keeping the relative flow rates through both pathways constant (Godard 2015). The GAP level at a temperature of 15 °C, however, exhibited a disproportionate decrease despite the constant flux ratios (Figure 5.31), hence resulting in a lowered Gibbs energy for the conversion of FBP to GAP and DHAP. The response of *B. megaterium* to cultivations at 45 °C differs greatly. Flux analysis indicated that the flux towards glycogen is reduced and peptidoglycan synthesis remains unchanged, thus more carbon is flowing through the central carbon metabolism. A closer look at the metabolite concentrations of FBP, DHAP, GAP and 3PG shows that all concentrations were lower than under optimal conditions (37 °C, 0M), while the PEP pool was increased. These findings could be a consequence of an inhibition of the phosphofructokinase by high concentrations of PEP shown in an earlier publication (Kelly & Turner 1969). Inhibition by thermodynamic constraints could be excluded, as the calculated Gibbs energies were equally low as the ones computed for the cultivation at 37 °C (Table 5.14).

Explanation of the behavior of the PYR pool is somewhat more complicated, as PYR is an important branching point, which is connected to the TCA cycle by pyruvate dehydrogenase and the anaplerotic reactions. Flux analysis of cultivations performed at 15 °C showed that the synthesis of all amino acids, with the exception of glutamine and glutamic acid, is reduced at low temperatures. This could result in a rising AEC, as a consequence of an elevated usage of the TCA cycle and thus an enhanced ATP synthesis. However, an AEC value 0.9 or higher would result in a serious down regulation of ATP-regenerating reactions, while at the same time up regulating ATP-utilizing reactions (Atkinson 1968). In order to maintain the AEC in the optimal range, cells use futile cycles to fine tune their energy level (Qian & Beard 2006). For *B. megaterium* it could be shown that the flux through malic enzyme, PYR-carboxylase and PEP-carboxykinase was increased by more than 50 % at a temperature of 15 °C (Godard 2015). Thus, the microorganism seemed to use anaplerotic reactions to regulate its AEC and to reduce the excess carbon of the TCA cycle by redirecting the flux towards PYR and PEP. Redirection of carbon combined with reduced synthesis of amino acids results in the disproportionate increase of the intracellular PYR concentration (Figure 5.31). In an attempt to reduce the amount of carbon, which enters the TCA cycle, the formation of biomass from PEP (Godard 2015) and secretion of PYR into the surrounding medium were increased. Both measures could not completely compensate the increased carbon influx, thus leading to lower driving force for the conversion of PEP to PYR. Additionally, disproportionately high amounts of AcCoA, CIT, ISOCIT and αKG have been measured, which might result from the reduced amino acid synthesis (Figure 5.31). The intracellular concentration of SUC at a temperature of 15 °C, however, was found to be ten times lower than the SUC pool measured for cultivations at optimal conditions (37 °C). Here, the increased synthesis of glutamic acid and biomass from αKG (Godard 2015) and the high amount of extracellular αKG measured via LC-MS/MS seem to function as an effective carbon sink. Reaction of the metabolic profile to temperatures of 45 °C was less extreme if compared to changes at 15 °C. Flux analysis had shown that the relative flux through the TCA cycle remained unchanged even though the amino acid synthesis was again reduced. Only an increase of the malic enzyme activity was monitored, as a consequence of the backlog from OA, which is then converted to AcCoA and subsequently secreted as acetate. In contrast to cultivations at low temperatures the excess carbon is used to generate more energy instead of plain secretion. Moreover, measurements of the metabolites of the TCA cycle revealed smaller intracellular concentrations of SUC and αKG than expected, while the extracellular amounts were increasing. As a result, the mass action ratios favored the formation of αKG, while the conversion of SUC to MAL was lowered. This phenomenon seemed an effective measure to reduce the stress of excess carbon resulting from the lowered anabolic reactions rates.

In summary, it can be noted that the combination of flux analysis, quantitative metabolomics and thermodynamics was successfully applied to elucidate effects of temperature induced stress on the central carbon metabolism of *B. megaterium*. It could be shown that extreme temperatures led to lowered growth rates accompanied by a reduced synthesis of amino acids. The resulting carbon backlog led to a perturbation of metabolite pools and thereby altered the existing driving forces. However, in good agreement with the response of *B. subtilis* to salt stress (Kohlstedt 2014), it could be shown that B. megaterium tries to maintain the intracellular metabolite ratios at all costs. Here, it could be shown that the distribution of relative fluxes through EMP- and PP pathway remains stable, while fluxes of the TCA cycle, secretion of organic acids and biomass generation showed a strong reaction to stress (Godard 2015). At a temperature of 15 °C *B. megaterium* used anaplerotic reactions in combination with biomass formation and increased secretion of PYR and αKG to stabilize the energetic state and the metabolite ratios. At 45 °C *B. megaterium* showed a higher growth rate, thus the carbon excess, due to reduced anabolic reaction rates, was smaller and the energy demand higher. Consequently, *B. megaterium* used the secretion of αKG and SUC to balance the metabolite profiles and converted AcCoA to acetate to supply additional energy, while simultaneously reducing the carbon influx into the TCA cycle.

5.4.3 Response of *B. megaterium* to Salt Induced Stress

The most obvious reaction of *B. megaterium* to the salt induced stress was the reduced growth rate in highly saline medium. The higher the salt concentration, the lower the measured growth rate (Table 5.17). It is known that the activity of the Glc-PTS in *E. coli* is linked to the PEP/PYR ratio (Hogema et al. 1998; Gabor et al. 2011), thus existence of a similar mechanism for *B. megaterium* seemed plausible. High PEP/PYR ratios indicate a low phosphorylation state of the Glc-PTS, which equals a low glucose phosphorylation rate. In contrast to the growth rate, measurements of the intracellular concentrations of PEP und PYR showed that the PEP/PYR ratio was slightly increased at a concentration of 0.6 M sodium chloride, but was strongly decreased at salt concentrations of 1.2 M (Table 5.17). Thus, the PEP/PYR ratio indicated towards a higher glucose uptake rate under highly saline conditions, while the growth rate was reduced under these conditions.

Table 5.17: Compilation of measured PEP/PYR ratios and growth rates exhibited by *B. megaterium* grown at optimal conditions (0M NaCl) and at stress inducing salt concentrations (0.6 M, 1.2 M). Cultivations were performed in baffled shake flasks (n=3), using minimal medium with glucose as carbon source. Metabolite samples were taken during exponential growth phase at a biomass of 1 g/L.

	0 M	0.6 M	1.2 M
PEP/PYR [-]	0.62	0.73	0.11
μ [h^{-1}]	1.19	0.69	0.39

Previous publications showed that salt stress induces lowered membrane fluidity in microorganisms, which hinders the glucose uptake rate (Hosono 1992; Los & Murata 2004). Hence, the lowered growth rate in highly saline media might be a result of lowered membrane fluidity and not a consequence of changes in the central carbon metabolism. Quite contrary, the low PEP/PYR ratio measured for cultivations in media containing 1.2 M sodium chloride indicates an enhanced phosphorylation of the Glc-PTS, in order to overcome the hindering fluidity. As discussed in chapter 5.3.2.1, the AEC value and the NADH/NAD ratio remained in optimal range despite the heightened salt concentrations. Conversely, a slightly decreased NADPH/NADP was measured in cells growing in medium containing 0.6 M sodium chloride and a strong increase of the ratio was detected for in medium containing 1.2 M sodium chloride (Table 5.15). Prokaryotes react very sensitive to substantial changes in the osmolarity caused by ions or organic solutes, as hypertonic conditions lead to dehydration while the cell bursts under hypotonic conditions. An initial response of most prokaryotes to salt induced stress is the enhanced *de novo* synthesis of proline, which functions as an osmoprotectant (Empadinhas & Da Costa 2008). In addition, *B. megaterium* is known to accumulate high levels of the storage polymer polyhydroxybutyrate (PHB) under the influence of high salt concentrations (Wang et al. 2005). Both responses demand high levels of carbon and NADPH, thereby explaining the distorted NADPH/NADP ratio. It could be seen that rising salt concentrations were accompanied by a strong increase of the total concentration of the redox pair NADPH and NADP, hereby indicating an increased demand of NADPH for synthesis of proline and PHB. However, the response barely compensated the demand of NADPH at the 0.6 M salt concentration and strongly over compensated it at high salt concentrations (Table 5.15). Hence, it seems that the cells had problems to fit the NADPH/NADP ratio to the actual demand at highly saline conditions.

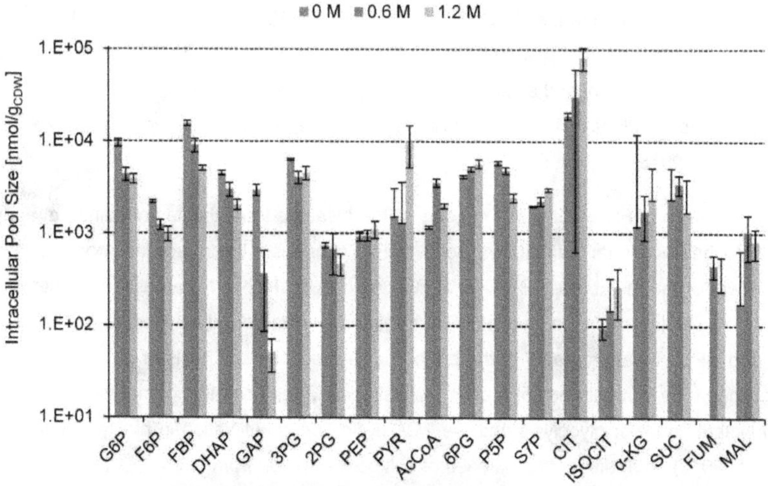

Figure 5.32: Effect of different salt concentrations (0 M, 0.6 M, 1.2 M) on metabolite pools of the central carbon metabolism measured in *B. megaterium*. *B. megaterium* was grown in shake flasks (n=3) containing modified M9 medium with glucose as carbon source. Metabolite samples were taken during exponential growth phase at a biomass of 1 g/L.

Combination of metabolite quantification, flux analysis and thermodynamics was then used to analyze the cellular response towards salt stress and to identify possible limitations and bottlenecks. Figure 5.32 depicts the intracellular concentrations of all metabolites of the central carbon metabolism at differing salinities of the medium. It becomes apparent that the pool size of metabolites of upper glycolysis and most metabolites of the pentose phosphate pathway decreased with rising salt concentration as a consequence of the diminished carbon uptake. The intracellular pools of 6PG and sedoheptulose 7-phosphate (S7P) on the other hand increased with rising salt concentration. It is likely, that the 6PG level was higher due to an elevated activity of glucose-6-phosphate dehydrogenase and 6-phosphogluconolactonase as a consequence of the higher NADPH demand. Accumulation of S7P on the other hand is not connected to a higher demand of redox equivalents or carbon at that point of the PP pathway. It seems more likely, that S7P accumulates because of the especially low GAP pool at high salt concentrations. The transaldolase reaction needs equal amounts of GAP and S7P to transform them to E4P and F6P. If one of the educts is missing, the other will accumulate as the reaction comes to a halt. A closer look into the calculated Gibbs energies confirms these observations. It can be seen that the extreme reduction of the GAP pool, the crucial connection of glycolysis and PP pathway, led to an increasing driving force for GAP forming reactions and simultaneously lowered driving forces for GAP consuming steps (Table 5.14). These findings were in good agreement with the performed flux analysis, which showed a reduced formation of glycogen, nucleotides and amino acids as a consequence of the reduced carbon uptake and an increased flow through the PP pathway to raise the NADPH/NADP ratio at both salt concentrations. Moreover, the synthesis of peptidoglycan from G6P and F6P, formation of biomass from the GAP pool and even the flux through glucose-6-phosphate isomerase were reduced at concentrations of 1.2 M sodium chloride, thereby enforcing flux through the PP pathway (Godard 2015).

As can be seen in Figure 5.32 the majority of the measured metabolites remained at a level similar to the level under optimal conditions. Drastic changes in the intracellular pool size could be shown for the metabolites PYR, GAP, MAL and FUM. Hence it was assumed that these pools play a major role in the reaction of *B. megaterium* towards salt induced stress. Flux analysis and HPLC measurements of the broth showed reduced synthesis of amino acids, with exception of the osmoprotectant proline, as well as lowered secretion of organic acids, apart from lactate, in lower glycolysis and the TCA cycle during cultivation in highly saline media (Godard 2015). In addition, the increase of salt concentration from 0.6 M to 1.2 M resulted in a lowered flux through anaplerotic reactions of malic enzyme and PEP-carboxykinase and reduced biomass formation from PEP and αKG (Godard 2015). Combination of the reduced rate of anaplerotic reactions and the lowered rates of anabolic reaction could explain the drastically higher levels of MAL and especially FUM. Reduction of the biomass formation from PEP combined with the lowered rate of amino acid synthesis from PYR led to a 10-fold increase of the intracellular PYR concentration at the highest salinity (Figure 5.32). The excess PYR was then used to form lactate. Thereby, maintaining the extremely low NADH/NAD ratio (Table 5.15), while simultaneously reducing the intracellular level of PYR. A high concentration of NAD leads to higher driving forces for NAD dependent reactions, thus facilitating glycolysis and the TCA cycle and here especially the conversion of PYR to AcCoA. The surplus of carbon was thus redirected into the metabolites AcCoA, CIT, ISOCIT up to the branching point αKG at which it was used for the enhanced formation of proline. Consequently, the metabolite pools increased with increasing salinity of the medium (Figure 5.32). A closer look at the metabolite pools of AcCoA, SUC, FUM and MAL shows, that the intracellular accumulation of these metabolites was higher in cells which were cultivated at a salinity of 0.6 M sodium chloride and not at 1.2 M. This is probably a consequence of the previously discussed NADPH/NADP ratio, which was the lowest in cells cultivated in 0.6 M medium. NADPH is needed during the synthesis of PHB from AcCoA (Dawes 1988) and during synthesis of proline from αKG (Pedersen et al. 1999). Consequently, the low availability of NADPH is followed by declining rates of PHB and proline formation. The resulting reduction of the carbon drain subsequently leads to an accumulation of the follow up reactions. In respect of proline and PHB formation, it has to be noted that *B. megaterium* seems to put the emphasis on generation of the compatible solute proline instead of the storage protein PHB. Cultivation of the microorganism in 0.6 M medium resulted in a 30-fold increase of proline and 15 % of the biomass being PHB. A further doubling of the osmolarity from 0.6 M to 1.2 M led to a smaller increase of proline but a doubling of the PHB proportion (Table 5.18). This setting of priorities seems reasonable, as proline ensures the immediate survival of the cell in its struggle with osmolarity, while the storage protein PHB becomes only important in the case of glucose depletion.

Table 5.18: Effect of salt induced stress on the intracellular accumulation of the osmoprotectant proline and the storage polymer PHB by *B. megaterium*. Cultivations were performed in baffled shake flasks (n=3), using minimal medium with glucose as carbon source. Metabolite samples were taken during exponential growth phase at a biomass of 1 g/L.

	0 M	0.6 M	1.2 M
Proline [µmol/g_{CDW}]	34	922	1339
PHB [%]	-	15	30

Summarizing, it can be stated that the combination of flux analysis, quantitative metabolomics and thermodynamics proved to be a powerful tool for the identification of changes in the central carbon metabolism of *B. megaterium* as a reaction towards stress induced by high salinity. It could be shown that extreme salt concentrations led to reduced growth rates which were accompanied by lowered rates for anabolic reactions. Higher salt concentrations resulted in lower anabolic activities of the cell. The excess carbon was redirected into the synthesis of the compatible solute proline and the storage protein PHB, still a massive accumulation of PYR could be identified. Furthermore, it was shown that *B. megaterium* uses lactate formation to maintain its NADH/NAD ratio and enforces the utilization of PP pathway to generate high amount of NADPH, which is needed in the formation of PHB and proline. The NADPH/NADP ratio was moreover identified to limit the generation of PHB and proline at salt concentrations of 0.6 M. At higher salt concentrations (1.2 M) the NADPH content was disproportionately high, while activity of the non-oxidative PP pathway seemed to be controlled by the availability of GAP. Both findings indicate dysregulation of *B. megaterium* in highly saline environments.

5.5 Impact of Nutrient Levels on the Central Carbon Metabolism of *D. shibae*

The previous chapters have shown that metabolite quantification combined with thermodynamics is a powerful tool for the elucidation of consequences of genetic modifications or environmental changes. Therefore, the method was used to investigate the effect of differing carbon sources on the central carbon metabolism of the marine organism *D. shibae*. Cultures of the microorganism were grown on the carbon sources succinate or glucose to enforce significant changes of metabolite levels. The results were then compared to metabolite levels of *P. putida* and *R. palustris* which were grown on glucose and succinate, respectively. Both microorganisms were used due to the similarity of their core metabolism under the applied conditions. *D. shibae* and *P. putida* degrade glucose exclusively via the ED pathway, while *R. palustris* utilizes succinate as the favorable carbon source.

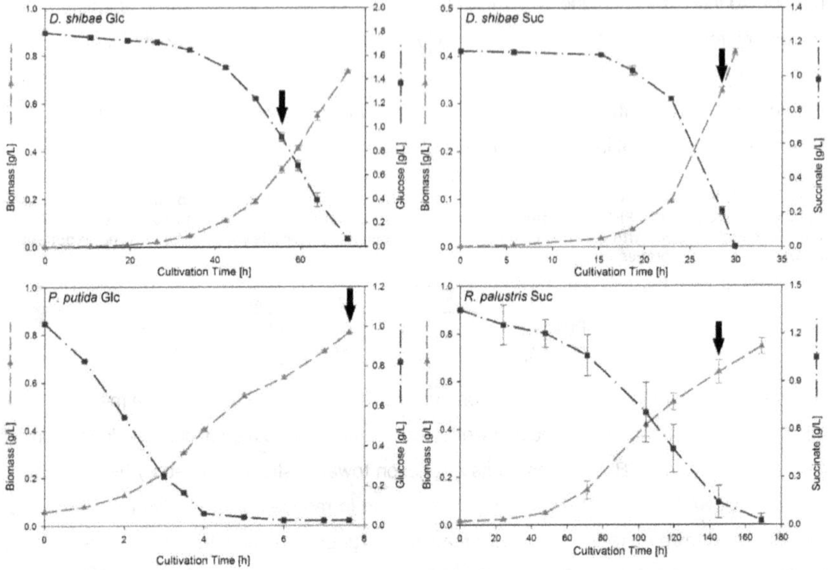

Figure 5.33: Cultivation profiles of *D. shibae*, *P. putida* and *R. palustris* grown in the corresponding minimal media. *D. shibae* was cultivated in shake flasks (n=3) using glucose or succinate as the sole carbon source. *P. putida* was grown in a 1L DASGIP reactor in fed-batch mode to minimize the formation of secondary metabolites. After the batch phase an exponential feed profile was applied, which supported growth at a rate of μ=0.15 h^{-1}. Samples were taken when the desired biomass of 0.3–0.8 g/L had been reached. *R. palustris* was cultivated under photoheterotrophic conditions, therefore anaerobic bottles were used (n=3).

Cultivations of *R. palustris* were performed in anaerobic bottles, while *D. shibae* was cultivated in shake flasks. *P. putida* was grown in a bioreactor using an exponential feed profile with a lowered growth rate to avoid overflow metabolism. The Cultivations of *D. shibae* exhibited exponential growth for both carbon sources used in the experiment and the exponential feed profile used for *P. putida* resulted in the desired growth rate of μ=0.15 h^{-1} during the feeding phase. The growth rate of *R. palustris* declined towards the end of the cultivation. As *R. palustris* grew under photoheterotrophic conditions, it used light to generate energy. However, with increasing biomass less light became available in the core of the anaerobic bottles. The self-shading effect could explain the observed decrease in growth rate with increasing biomass. *D. shibae* was sampled at a cellular dry weight of 0.3 g/L, *R. palustris* at a cellular dry weight of 0.6 g/L and *P. putida* at a dry weight of 0.8 g/L (n=3). Obviously, enough substrate was left in the surrounding medium at the time of sampling, as all cultures continued growing afterwards (Figure 5.33). Hence, a corruption of metabolic datasets, due to depleted carbon pools, could be excluded.

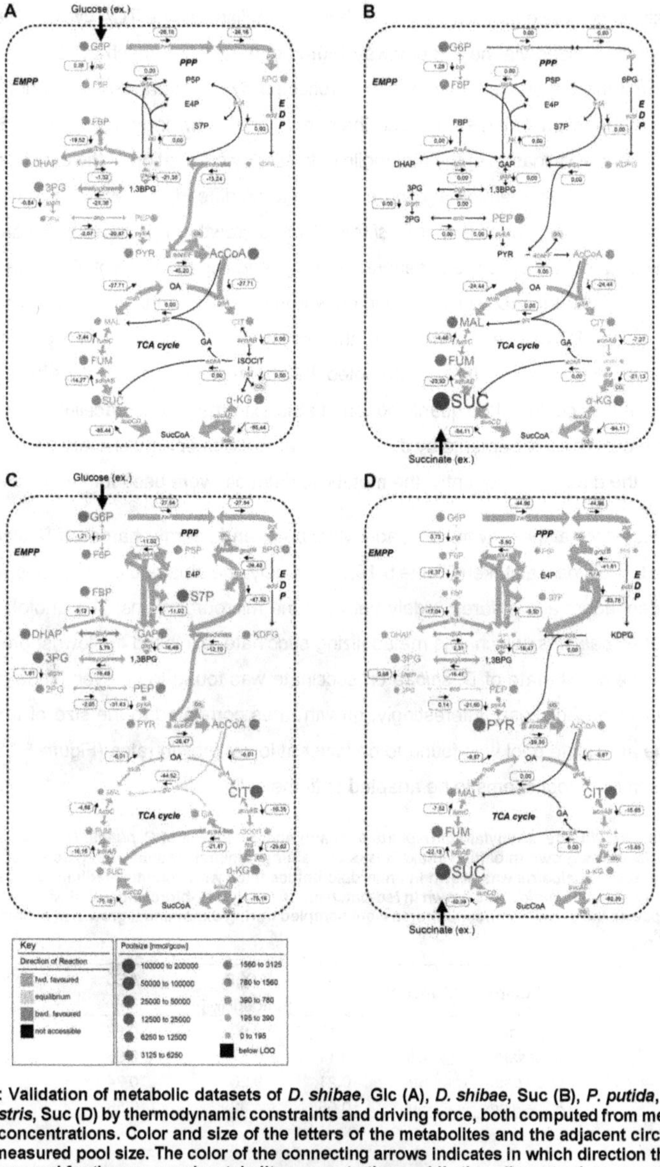

Figure 5.34: Validation of metabolic datasets of *D. shibae*, Glc (A), *D. shibae*, Suc (B), *P. putida*, Glc (C) and *R. palustris*, Suc (D) by thermodynamic constraints and driving force, both computed from measured metabolite concentrations. Color and size of the letters of the metabolites and the adjacent circles represent the measured pool size. The color of the connecting arrows indicates in which direction the reaction would proceed for the measured metabolite concentrations, while the adjacent values represent the computed Gibbs energies for the corresponding reaction. Negative Gibbs energies support a forward directed reaction, positive energies result in a back reaction. *D. shibae* was grown in baffled shake flasks containing minimal medium with glucose or succinate as carbon source. *R. palustris* was grown in anaerobic bottles filled with minimal medium, using succinate as the carbon source. *P. putida* was grown in fed-batch mode using a 1L-bioreactor filled with minimal medium and glucose as the carbon source. Cultures were sampled during exponential growth at a biomass in range of 0.3-0.8 g/L. GA and KDPG could not be found in D. shibae and R. palustris, due to the inactivity of the ED pathway and the GA cycle. Consequently, no driving forces were calculated for reactions using these two metabolites.

Obviously, succinate enters the central carbon metabolism via the TCA cycle, while glucose is primarily metabolized via the ED pathway (Fürch et al. 2009; Bartsch 2015). Thus, it was investigated if the forced gluconeogenesis is reflected by the metabolite pools of the core metabolism of the investigated strains and the corresponding driving forces. Figure 5.34 depicts the results of the validation of the metabolic data sets by calculation of the corresponding driving forces and Gibbs energies. It can be seen that the differing substrates resulted in drastic changes of the core metabolism of *D. shibae*. During growth on glucose *D. shibae* utilized the ED pathway and the TCA cycle to generate energy. Additionally, parts of glycolysis were found to be active. Cultures of *D. shibae*, which grew with succinate as the carbon source, exhibited a highly active TCA cycle. Furthermore, the computed driving forces indicated towards the formation of G6P from F6P. It has to be noted that the majority of metabolites of glycolysis, ED and PP pathway could not be quantified, due to the extremely low intracellular concentrations. Validation of the measurements by driving force and Gibbs energy showed no proof of a corruption of the data. Consequently, the metabolic datasets were used for further analysis.

First a closer look at the growth rates, adenylate pools and energy charges of *D. shibae*, *R. palustris* and *P. putida* was taken (Table 5.19). Obviously, the obtained growth rates were carbon source dependent and differed widely between the microorganisms. The photoheterotrophic growth of *R. palustris*, which was metabolizing succinate, exhibited the lowest growth rate. In contrast, the growth rate of *D. shibae* on succinate was found to be even higher than during metabolization of glucose. Interestingly, growth rates correlated to the size of the adenylate pool. The adenylate pool was found to be lower at lower growth rates (Figure 5.35). The size of the adenylate pool seems to be adapted to fit the cellular needs.

Table 5.19: Growth rate, adenylate phosphate pool and energy charge of *P. putida*, *R. palustris* and *D. shibae*. *D. shibae* was grown in baffled shake flasks containing minimal medium with glucose or succinate as carbon source. *R. palustris* was grown in anaerobic bottles filled with minimal medium, using succinate as the carbon source. *P. putida* was grown in fed-batch mode using a 1L-bioreactor filled with minimal medium and glucose as the carbon source. Cultures were sampled during exponential growth at a biomass between 0.3-0.8 g/L.

Strain	Substrate	μ [h^{-1}]	A(X)P [μmol/g_{CDW}]	AEC [-]
P. putida	Glc	0.15	7.92	0.62
D. shibae	Glc	0.11	6.51	0.86
D. shibae	Suc	0.21	9.26	0.74
R. palustris	Suc	0.05	4.32	0.56

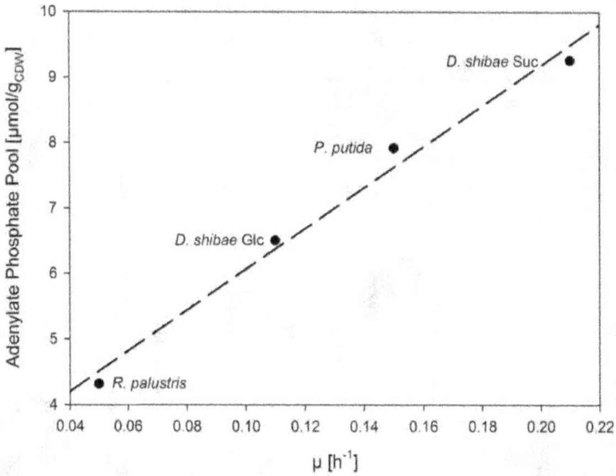

Figure 5.35: Correlation of adenylate phosphate pool size and growth rate. Measurements of the A(X)P pool size were obtained from aerobic cultures of *D. shibae* with succinate or glucose as the carbon source, as well as anaerobic cultures of *R. palustris* with succinate. Cultures of *P. putida* were grown in fed-batch mode with glucose as the sole carbon source. Cultures were sampled during exponential growth at a biomass between 0.3-0.8 g/L.

Analysis of the energy charge showed that the AEC of *D. shibae* was well in the range of viable cells predicted by Atkinson (1968), thereby confirming the preceding validation steps. Even though the energy charge of *P. putida* was below the threshold of 0.7, the value was in good agreement with previous data of fed-batch cultivations (Jenkins et al. 1987).

In a next step the metabolite pools of the cultivations of *P. putida* and *D. shibae*, both using glucose as the sole carbon source, were compared to detect differences in the central carbon metabolism of both microorganisms (Figure 5.36). It can be seen that levels of glycolysis and ED pathway metabolites were higher for *P. putida*, while most metabolite levels of the TCA cycle were found to be higher in *D. shibae*. Furthermore, it could be shown that *P. putida* tends to accumulate intracellular CIT, while *D. shibae* accumulated FUM and SUC.

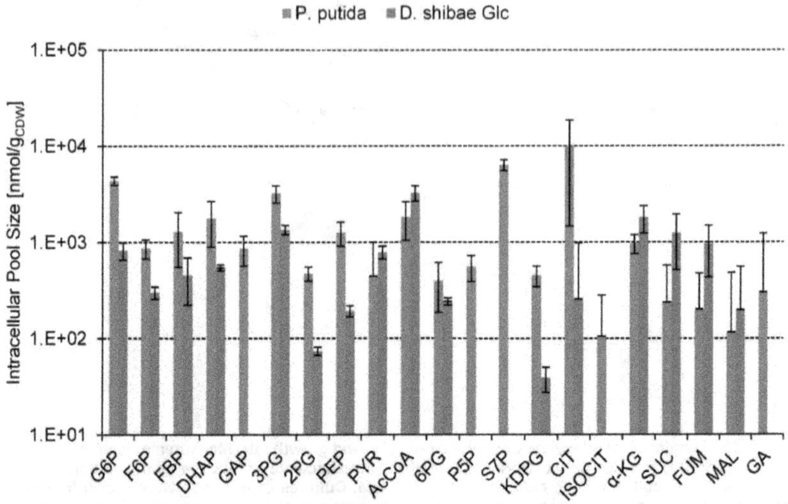

Figure 5.36: Comparison of intracellular metabolite pools of the central carbon metabolism of *P. putida* and *D. shibae* (n=3). *D. shibae* was grown in a baffled shake flask, while *P. putida* was cultivated in a 1L-bioreactor operated in fed-batch mode to minimize formation of secondary metabolites. Both microorganisms were grown in their specific minimal medium with glucose as carbon source. Metabolite samples were taken at a biomass of 0.3 g/L (*D. shibae*) and 0.8 g/L (*P. putida*).

The most remarkable difference between both microorganisms was the lack of PP pathway metabolites in *D. shibae* compared to the substantial amounts of P5P and S7P in *P. putida*. These results are supported by the publication of Klingner et al. (2015), stating that metabolic flux through the PP pathway could be shown for *P. putida* but not for *D. shibae*.

Figure 5.37 depicts the comparison of the intracellular metabolite pools of *D. shibae* and *R. palustris* growing with succinate as the carbon source. It can be seen that all metabolites of glycolysis, PP pathway and TCA cycle were present in *R. palustris* under the applied conditions. Measurements of *D. shibae* on the other hand showed, that most metabolites of glycolysis and the PP pathway were below the detection limit. G6P, F6P, PEP and KDPG were the only detectable metabolites. This could indicate that *D. shibae* uses these pathways only to a minor extent, i.e. for anabolic purposes.

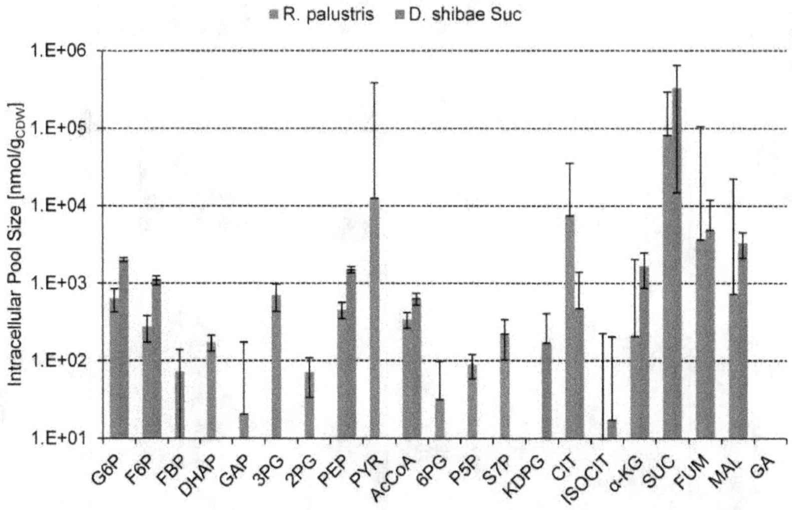

Figure 5.37: Comparison of intracellular metabolite pools of the central carbon metabolism of *R. palustris* and *D. shibae* (n=3). *D. shibae* was grown in a baffled shake flask, while *R. palustris* was cultivated in anaerobic bottles to enable photoheterotrophic growth. Both microorganisms were grown in their specific minimal medium with succinate as carbon source. Metabolite samples were taken at a biomass of 0.3 g/L (*D. shibae*) and 0.5 g/L (*R. palustris*).

As a consequence, it was assumed that both microorganisms employ a fully functional TCA cycle if growing on succinate. The presence of detectable amounts of metabolites of PP pathway and glycolysis in *R. palustris* could be the consequence of an active Calvin-Benson-Bassham cycle, which is used to provide precursors and to regenerate cofactors under anaerobic conditions (McKinlay & Harwood 2010).

Summarizing, the change of carbon sources provoked drastic changes in the central carbon metabolism of *D. shibae*. Comparison of the metabolic datasets of *D. shibae* to *P. putida* and *R. palustris* revealed large differences as a consequence of differing pathway utilization, even though the same carbon sources were metabolized.

5.6 Integrated Analysis of Metabolic Datasets of Different Microbial Strains

In a final step of exploration, all datasets gathered during the experiments were now integrated. Generally, the measured metabolite levels were found feasible on basis of the underlying thermodynamics (Table 5.20). In almost all cases the calculated Gibb's energies complied with the observed flux direction. This confirms the validity of the developed approach across all involved steps from sampling to the final data evaluation.

Table 5.20: Comparative list of Gibbs energies calculated from metabolite pools of seven different microbial strains by application of Equation 3.2. All microorganisms were grown in their specific minimal medium with glucose as the carbon source. An exception was *R. palustris*; here succinate was used as the carbon source. Metabolite samples (n=3) were taken during exponential growth of the microorganisms and during carbon excess. Standard Gibbs energies were obtained from the database for thermodynamic properties (Li et al. 2011). Gibbs energies supporting a forward directed reaction are highlighted in green, while a red background equals a backward directed reaction.

Pathway	Reaction	*C. glutamicum*	*E. coli* K12	*B. megaterium*	*Y. pseudotuberculosis*	*D. shibae*	*R. palustris*	*B. subtilis*	*P. putida*
EMP	G6P → F6P	-1.38	-1.58	-0.84	-2.36	0.28	-0.49	-2.71	-1.21
	F6P → FBP	-7.99	-13.01	-13.04	-5.44	-17.43	-16.61	-4.49	-12.95
	FBP → GAP + DHAP	-21.90	-13.24	-9.93	-17.74	-19.52	-7.90	-12.70	-9.12
	DHAP → GAP	-4.46	3.44	6.53	0.81	-1.32	9.10	3.99	5.79
	GAP → 3PG	-9.50	-15.49	-26.85	-20.91	-21.38	-20.20	-23.76	-16.49
	3PG → 2PG	-3.38	1.79	1.02	1.27	-0.84	1.89	2.59	1.61
	2PG → PEP	-2.88	-1.65	-3.89	-5.81	-2.07	-1.73	-5.32	-2.05
	PEP → PYR	-23.93	-23.61	-23.73	-14.22	-20.87	-32.28	-26.31	-31.43
	PYR → AcCoA	-28.33	-34.48	-42.43	-46.76	-45.20	-30.79	-32.49	-26.47
ED	G6P → 6PG	-25.04	-27.46	-25.58	-25.65	-26.16	-31.49	-29.92	-27.84
	KDPG → GAP + PYR	-	-	-	-	-15.24	-	-8.29	-12.70
PP	6PG → RIBU5P	-27.46	-29.82	-28.59	-33.61	-	-27.88	-32.12	-26.40
	2 6PG → GAP + S7P	-62.70	-53.82	-62.15	-73.07	-	-49.07	-65.15	-47.52
	GAP + S7P → F6P + GAP	-8.20	-6.42	-5.80	-6.74	-	-6.16	-6.59	-11.02
TCA cycle	MAL → AcCoA+ OA → CIT	-11.34	-18.90	-12.36	-31.93	-27.71	-11.65	-8.13	-0.01
	CIT → ISOCIT	-4.47	-	-12.16	0.92	-	-6.21	-8.58	-10.35
	ISOCIT → AKG	-24.91	-	-26.50	-34.01	-	-29.14	-36.96	-25.62
	AKG → SUC	-80.41	-75.44	-79.03	-71.51	-85.44	-65.37	-72.91	-75.19
	SUC → FUM	-6.66	-12.89	-	-14.32	-14.27	-14.62	-13.79	-16.16
	FUM → MAL	-6.51	0.39	-	3.44	-7.44	-14.42	-2.33	-4.88
GA-Shunt	ISOCIT → GA + SUC	-	-	-	-	-	-	-	-21.47
	GA + AcCoA → MAL	-	-	-	-	-	-	-	-44.52

One should, however, notice that this does not hold for the reaction of triose phosphate isomerase (Table 5.20). It is known, that a large excess of DHAP is needed to drive the synthesis of GAP. The ratio between the two metabolites has been estimated as 96:4 (Cornish-Bowden 1981). The observed ratio between DHAP and GAP was high enough to drive the glycolytic flux in *C. glutamicum* and glucose grown *D. shibae* (Wittmann & Heinzle 2002; Bartsch 2015) and within a certain error range also appeared acceptable for *Y. pseudotuberculosis* (Bücker 2014). It also matched with the gluconeogenic flux of *P. putida*, lacking a functional glycolysis (Nikel et al. 2015), and of succinate grown *R. palustris*, employing the Calvin-Benson-Bassham cycle and gluconeogenesis to supply biomass precursors (Joshi et al. 2009).

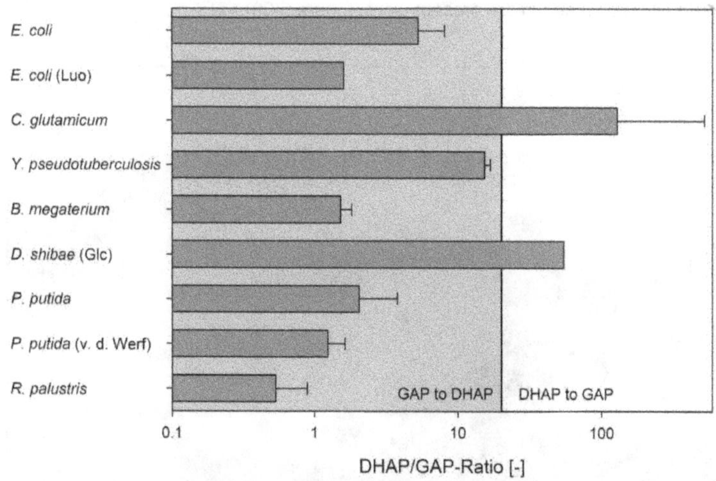

Figure 5.38: Intracellular DHAP/GAP ratios exhibited by seven microbial strains, which were acquired in the course of this work. For comparison ratios computed from metabolic datasets of previous publications were added for *E. coli* (Luo et al. 2007)and *P. putida* (van der Werf et al. 2008). Most strains were grown with glucose, while *R. palustris* was cultivated using succinate as the carbon source. Metabolite samples (n=3) were taken during exponential growth of the microorganisms and during carbon excess. The grey area equals ratios for which the formation of DHAP from GAP is favored. Higher ratios favor the formation of GAP from DHAP.

However, a strong mismatch was observed for *E. coli* and *B. megaterium*. Here, the metabolite ratio suggested a backward conversion of GAP into DHAP, which did not agree with the opposite direction of flux, inferred from ^{13}C flux data (Wittmann et al. 2007; Godard 2015). The exact reason remains unclear. Possible limitations in the analytics with a rather broad and shallow peak for GAP, which was a bit difficult to clearly separate from background noise, can only partly serve as an explanation. The DHAP/GAP ratio of *B. megaterium* was simply too low to be the result of a simple overestimation of the GAP pool. Interestingly, the comparison of the data with previous metabolome data (Luo et al. 2007; van der Werf et al. 2008) revealed that these exhibit an even higher deviation from what is expected on the basis of molecular flux direction (Figure 5.38). To which extent this is due to a specific phenomenon around triose phosphate isomerase, such as metabolite channeling, which could explain such an apparent mismatch, remains to be elucidated. In addition, it can be noticed that most glycolytic reactions operated close to their equilibrium, exceptions being reactions such as those catalyzed by phosphofructokinase and pyruvate kinase. This enables the cell to quickly rearrange flux upon perturbations and agrees well with our current view on this part of carbon core metabolism (Becker & Wittmann 2015).

Furthermore, a principle component analysis (PCA) was performed to statistically analyze the metabolic datasets obtained in the present work (Figure 5.39).

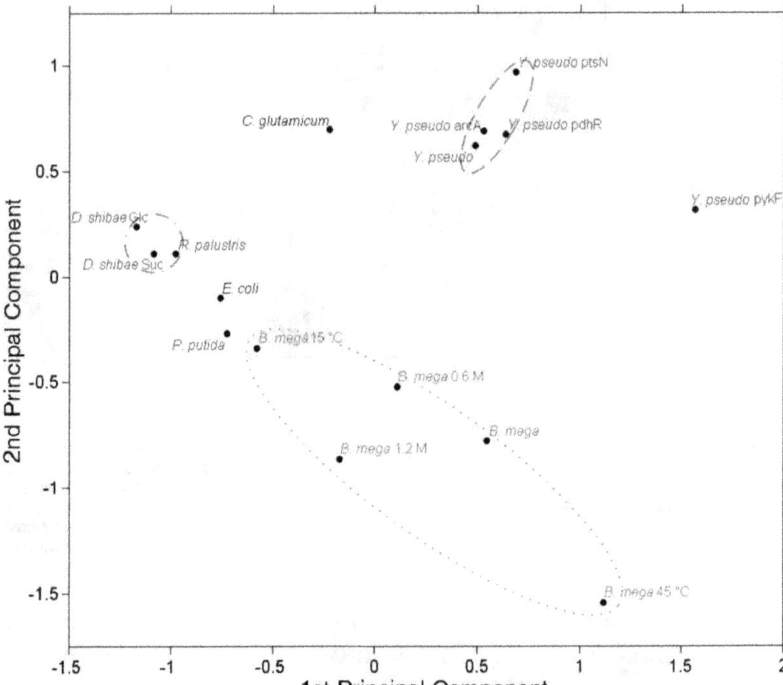

Figure 5.39: Principle Component Analysis (PCA) of the central carbon metabolome of different strains obtained in the present work. Metabolome datasets were generated during method validation (black), investigation of mutants of *Y. pseudotuberculosis* (red), stress inducing cultivations of *B. megaterium* (green) and during cultivations of *D. shibae* using different carbon sources (blue). Metabolite samples (n=3) were taken during exponential growth of the microorganisms and during carbon excess.

For this purpose, the metabolite levels were standardized (using R), as concentrations differed by several orders of magnitudes. The PCA in R then allowed evaluating if intracellular metabolite inventories were alike and to which extent genomic alterations and stress conditions led to change in the metabolome signature.

The PCA resulted in the formation of two distinct and one larger cluster. The core metabolism of *Y. pseudotuberculosis* and three of its four mutant strains were highly homologous and formed a cluster. The metabolome of the ΔpykF strain differed greatly, as the PCA shows a clear spatial separation. This is interesting to note, as this genetic modification was only rather local, whereas the other mutants lacked global regulators. The second cluster consisted of the metabolome of *R. palustris* and *D. shibae*, despite differences in the pathway usage and carbon source. A possible explanation is the enhanced usage of the TCA cycle, which was found in all three cultivations. Interestingly, salt stress affected the metabolome of *B. megaterium* to a lesser extent than temperature stress. The relatively conserved metabolic fingerprint matches with the recent discovery that *B. subtilis* maintains flux homeostasis at elevated salt stress (Kohlstedt et al. 2014). *C. glutamicum* differed from all other organisms and conditions and obviously exhibits an individual type of metabolism regarding flux and regulation.

6 Conclusions and Outlook

In the course of this work an extremely powerful method for quantification of metabolites of the central carbon metabolism was developed and validated, which was then applied to analyze the effects of genetic modification, stress, different substrates or cultivation techniques on the central carbon metabolism of various bacterial strains. Evaluation of the gained metabolomic datasets was performed by thermodynamic and energetic considerations. Combination of metabolomics and fluxomics approaches enabled a holistic and systems-oriented view during analysis of the reaction networks.

The novel quantification technique allowed a holistic analysis of the reaction networks by offering a considerable reduction of time per measurement. The drastic reduction of acquisition time, from more than 90 minutes to 25 minutes, by utilization of ion-pairing reagents and UHPLC enables a higher sample throughput. Thorough validation of existing techniques in combination with a set of novel efficient tools resulted in a high throughput method, which is able to quantify key metabolites involved in the major pathways of central carbon metabolism with a high dynamic range, down to femtomolar quantities, despite the presence of various biological matrices.

Concerning extraction protocols, it was found that cold methanol quenching allowed efficient quenching of all metabolic activities during the sampling process, thereby preventing falsification of metabolic data. Problems of metabolite leakage during the quenching process were successfully avoided by application of the differential method. With boiling ethanol and acidic acetonitrile methanol two extraction protocols were evaluated for their extraction efficiency. Boiling ethanol showed the higher extraction efficacy. In addition, the boiling ethanol showed a higher reproducibility, due to immediate inactivation of all enzymes at high temperatures. Furthermore, extraction with boiling ethanol consumed less time per extraction and was thus selected as the method of choice for the high throughput approach.

Evaluation of the metabolic dataset by calculation of adenylate energy charge, driving force and Gibbs energy, computed from the measured metabolite pools, allowed identification of datasets within a physiological meaningful range. These evaluations showed the high quality of the obtained datasets, as all metabolite pools supported ongoing activity of the catabolic pathways.

As a first application, quantitative metabolite measurements were used to investigate the energy metabolism of *E. coli* under various conditions. It could be shown that the energy charge of *E. coli* drops during times of glucose deprivation. To maintain the AEC within the optimal range of 0.7 to 0.9 large amounts of AMP and ADP accumulated in the medium, thereby lowering the overall concentration of intracellular adenylate phosphates but simultaneously increasing the energy charge. Furthermore, it could be shown by analysis of chemostat cultures, that the growth rate is proportional to the adenylate phosphate pool, while the energy charge was independent. However, inhibition of the respiratory chain by dinitrophenol led to a measurable decrease of growth rate and energy charge. These observations demonstrate that microorganisms have effective mechanisms at their disposal to maintain intracellular metabolite ratios at an optimal level. Only serious distortions of energy generation rates resulted in critical intracellular energy levels.

Furthermore, metabolite quantification was applied to investigate the reaction of *E. coli* to an insufficient supply of carbon, as it can be found during fed-batch cultivations. Here, the method was extended to detect all measurable isotopologues of the metabolites of the central carbon metabolism. By this extension it was possible to determine the isotopologue ratios throughout a fed-batch experiment which started with a batch phase using unlabeled glucose while during the feeding phase only fully labeled glucose was supplied. Consequently, it was possible to determine if the measured carbon originated from the feeding process or was part of earlier accumulated storage polymers. Analysis of the isotopologue ratios revealed, that *E. coli* uses all accessible carbon sources to maintain its optimal energy charge as long as possible. It could be shown that glycogen and amino acids, probably originating from protein degradation, were used to compensate the lack of carbon. In addition, it was shown that the cells remained growing with the growth rate intended by the applied feeding profile, while the energy charge dropped to a value of 0.4. The extremely low energy charge could be a consequence of the formation of subpopulations which oscillate between states of ATP utilization and regeneration, thus exhibiting an extremely low AEC while being perfectly viable at the same time. However, the long-term fed-batch experiments indicated that an AEC below 0.3 equals a dying population, as the AEC did not fall below this threshold despite the formation of oscillating subpopulations. This finding has to be kept in mind, when evaluating metabolic datasets of larger cell populations, in order to avoid false conclusions, as even a small subpopulation might result in serious variations of the average metabolite concentrations.

The new quantification method was further used to evaluate the impact of genetic modification on the central carbon metabolism of *Y. pseudotuberculosis* and thus the virulence of the pathogen. Comparison of metabolic datasets of *E. coli* and *Y. pseudotuberculosis* implicated that *Y. pseudotuberculosis* exhibits a lower energy charge, thereby enhancing glucose uptake and consequently PYR production. High titers of PYR provide protection against reactive oxygen species which are part of the antibacterial reaction of the mammalian immune system. For further investigation of the PYR accumulation in *Y. pseudotuberculosis*, one of the three regulators ArcA, PdhR and PtsN or the pyruvate kinase F were knocked out, all of which are tightly associated with regulation of the pyruvate node. Analysis of the obtained metabolic datasets allowed linking of reduced virulence with changes in the metabolome as a consequence of the genetic modifications. It could be shown that the deletion of ArcA results in a shift of fluxes from the energy efficient TCA towards rapid but less efficient fermentation. Additionally, it could be shown that the ΔpykF strain circumvented the deletion of pyruvate kinase by utilization of PEP carboxylase and malic enzyme to maintain a high PYR pool despite the high energy demand of these reactions. Both of these ATP intense modifications of the standard metabolism lead to a huge disadvantage of the two knockout strains ΔpykF and ΔArcA when competing for resources. The deletions of PtsN and PdhR resulted only in minor changes of metabolome and driving forces, therefore a link between virulence and metabolism could not be established. However, it has to be noted that the metabolic datasets were obtained from aerobically growing bacteria, which used glucose as the sole carbon source. Coinfection experiments have shown that PtsN and PdhR mutants exhibit a lowered virulence in liver and spleen, where glucose and oxygen are scarce. Hence, it is likely that metabolite quantification and flux analysis of *Y. pseudotuberculosis* strains, which were grown using succinate, arginine or ornithine as the sole carbon source, would give a more representative picture of the metabolic changes induced by the gene deletions.

Application of the new metabolite quantification method for the investigation of the stress response of *B. megaterium* to extreme temperatures or salt concentrations revealed the power of a combined metabolomics and fluxomics approach. Extreme temperatures led to lowered growth and amino acid synthesis, which resulted in perturbed metabolite pools and subsequently in altered driving forces and thermodynamics. It could be shown that *B. megaterium* balanced the carbon excess by enhanced secretion of organic acids and pyruvate to stabilize the intracellular driving forces. Simultaneously anaplerotic reactions were used for futile cycling during energy excess, while the enhanced conversion of AcCoA to acetate was utilized to generate additional energy. Both reactions to temperature induced stress allow *B. megaterium* to maintain an optimal energy level.

High salt concentrations also resulted in lowered anabolic reaction rates. However, the resulting carbon excess was redirected into the synthesis of the compatible solute proline and the storage protein PHB. Instead of the secretion of excess carbon, the metabolite analysis revealed an accumulation of carbon in the lower glycolysis. Furthermore, it was shown that *B. megaterium* uses lactate formation to maintain its NADH/NAD ratio and enforces the utilization of PP pathway to generate high amount of NADPH, which is needed in the formation of PHB and proline. The NADPH/NADP ratio was moreover identified to limit the generation of

PHB and proline at salt concentrations of 0.6 M. At higher salt concentrations (1.2 M) the NADPH content was disproportionately high, while activity of the reductive PP pathway was controlled by the availability of GAP. Both findings indicate dysregulation of *B. megaterium* in highly saline environments. Summarizing, it can be stated that the combination of flux analysis, quantitative metabolomics and thermodynamics proved to be a powerful tool for the identification of changes in the central carbon metabolism as a reaction towards stress.

Metabolite analysis of *D. shibae* growing on different substrates revealed the metabolic variability of the marine microorganism. It was shown that a change of the carbon source results in dramatic changes on metabolite level. During the growth on succinate metabolites of glycolysis, Entner-Doudoroff pathway or pentose phosphate pathway are almost completely absent, thus resulting in a measurable driving force in favor of gluconeogenesis (Figure 5.34 B). Furthermore, it could be shown that *D. shibae* employs a backward directed non-oxidative PP pathway to satisfy its needs for the precursor metabolites E4P and R5P. Despite these drastic changes, the energy charge was in the optimal range. Comparison of the metabolite levels of *D. shibae* to metabolite levels of glucose utilizing *P. putida* and succinate degrading *R. palustris* showed that the core metabolism of *D. shibae* is highly efficient at extremely low metabolite pools.

The power of the developed metabolite analysis tool was demonstrated by integrated analysis of all datasets gathered during the experiments. Generally, the measured metabolite levels resulted in Gibb's energies which complied with the flux direction estimated from ^{13}C flux data. Thus, measured metabolite levels seemed feasible on basis of the underlying laws of thermodynamics. The reaction catalyzed by triose-phosphate-isomerase was found to be an exception to the rule as the measured metabolite levels supported a flux opposite to the results of ^{13}C flux data. To which extent this is due to a specific phenomenon around triose phosphate isomerase, such as metabolite channeling, which could explain such an apparent mismatch, remains to be elucidated.

Statistical analysis of the metabolic datasets by means of PCA allowed evaluation of the influence of genomic alterations and stress onto the specific metabolome signatures of the investigated strains. Interestingly, the deletion of regulator proteins in *Y. pseudotuberculosis* showed a relatively small impact on the metabolic fingerprint in contrast to the deletion of the enzyme pyruvate kinase. Furthermore, it could be shown that environmental conditions have a serious impact on the metabolome. These results emphasize the potential of metabolome analysis in the investigation of microorganisms and their reaction to environmental changes and genomic alterations.

Summarizing it can be stated, that the newly established metabolite quantification method was successfully applied for investigations on the central carbon metabolism. It proved to be a resourceful method to gain insights and explain various impacts on the central metabolism, which could in the future be useful for various applications. Clinically, an improved understanding of metabolic changes in microorganisms due to genetic modifications or environmental conditions and the development and assessment of potential treatment measures based on these insights could prove extremely valuable in the future.

Moreover, the newly developed method is offering potential of high throughput screening, which could provide valuable insight into the metabolism of microorganisms in a wide area of application fields in system biotechnology, as it also easily defines metabolic bottlenecks. Hence, the newly established method has the potential to become a key technology in the optimization of biotechnological production processes.

7 Abbreviations and Symbols

Abbreviations

αKG	α-Ketoglutarate
1,3-BPG	1,3-Bisphsophoglycerate
2PG	2-Phosphoglcerate
3PG	3-Phosphoglycerate
6PG	6-Phosphogluconate
AA	Acetic acid
AAM	Acidic acetonitrile methanol
AAP	aerobic anoxygenic phototrophs
AcCoA	Acetyl-Coenzyme A
ADP	Adenosine diphosphate
AEC	Adenylate energy charge
AMP	Adenosine monophosphate
anNET	analytical network evaluation tool
ANP	Aqueous normal phase
APCI	Atmospheric pressure chemical ionization
ArcA	Anoxic redox control regulator
ATCC	American type culture collection
ATP	Adenosine triphosphate
BE	Boiling ethanol
CAD	Collision gas
CBB	Calvin-Benson-Bessham
CDW	Cellular dry weight
CE	Capillary electrophoresis
CE	Collision energy
CID	Collision induced dissociation
CIT	Citrate
CO_2	Carbon dioxide
CoA	Coenzyme A
CUR	Curtain gas
CV	Coefficient of variation
CXP	Cell exit potential
DHAP	Dihydroxyacetone phosphate
DIN	German institute of standardization
DNP	Dinitrophenol
DP	Declustering potential
DSMZ	German collection of microorganisms and cell cultures
E	Educts
E4P	Erythrose 4-phosphate
ED	Entner-Doudoroff
EI	Electron impact
EMP	Embden-Meyerhof-Parnas/glycolysis
EP	Entrance potential

ESI	Electrospray ionization
EtOH	Ethanol
F1P	Fructose 1-phosphate
F6P	Fructose 6-phosphate
FAD	Flavine adenine dinucleotide (oxidized)
$FADH_2$	Flavine adenine dinucleotide (reduced)
FBP	Fructose 1,6-bisphosphate
FT	Fourier transform
FUM	Fumarate
G6P	Glucose 6-phosphate
GA	Glyoxylic acid
GAP	Glyceraldehyde 3-phosphate
GC	Gas chromatography
GC-MS	Gas chromatography-mass spectrometry
Glc	Glucose
Glc-PTS	Glucose-phosphotransferase system
GS1	Nebulizer gas
GS2	Auxiliary gas
GTP	Guanosine triphosphate
HILIC	Hydrophilic interaction liquid chromatography
HPLC	High pressure liquid chromatography
IP-LC	Ion pairing liquid chromatography
IS	Ion spray voltage
ISOCIT	Isocitric acid
KDPG	2-Keto-3-deoxyphosphogluconate
LB	Luria-Bertani
LC	Liquid chromatography
LC-MS	Liquid chromatography-mass spectrometry
LC-MS/MS	Liquid chromatography-tandem masspectrometry
LOD	Limit of detection
LOQ	Limit of quantification
m/z-ratio	Mass to charge ratio
MAL	Malate
MAR	Mass action ratio
MeOH	Methanol
MRM	Multiple reaction monitoring
MS	Mass spectrometry
NAD	Nicotineamide adenine dinucleotide (oxidized)
NADH	Nicotineamide adenine dinucleotide (reduced)
NADP	Nicotineamide adenine dinucleotide phosphate (oxidized)
NADPH	Nicotineamide adenine dinucleotide phosphate (reduced)
OA	Oxaloacetate
OD	Optical density
P	Products
PDH	pyruvate dehydrogenase
PDHC	Pyruvate dehydrogenase complex

pdhR	Pyruvate dehydrogenase complex regulator
PEP	Phosphoenol pyruvate
PFPP	Pentafluorophenylpropyl
PHA	polyhydroxyalkanoates
PHB	polyhydroxybutyrate
P_i	Phosphor (inorganic)
PNB	purple non-sulfur bacteria
PP	Pentose phosphate
pykF	Pyruvate kinase F
PYR	Pyruvate
Q	Quadrupole
R5P	Ribose 5-phosphate
RIBU5P	Ribulose 5-phosphate
RP	Reversed phase
RP-IP	Reversed phase ion pairing
RT	Retention time
S7P	Seduheptulose 7-phosphate
SIM	Single ion monitoring
SUC	Succinate
SucCoA	Sucinnyl-Coenzyme A
TBA	Tributylamine
TCA	Tricarboxylic acid
TEM	Heat
TLC	Thin-layer chromatography
TOF	Time of flight
UHPLC	Ultra-high performance liquid chromatography
UV	Ultra violet
X5P	Xylulose 5-phosphate
XIC	Extracted ion chromatogram

Symbols

ΔG	Gibbs formation energy	[J/mol]
$\Delta G^{o'}$	Standard Gibbs energy	[J/mol]
μ	Specific growth rate	[h^{-1}]
D	Dilution rate	[h^{-1}]
F	Feeding rate	[mL/h]
Keq	Equilibrium constant	[mol/mol]
R	Ideal gas constant	[J/(molK)]
S	Substrate	[g/L]
T	Temperature	[°C]
t	Time	[h] or [min]
V	Volume	[L]
X	Biomass	[g/L]
Y X/S	Biomass yield	[g/g]

8 References

Akinterinwa, O., Khankal, R. & Cirino, P.C., 2008. Metabolic engineering for bioproduction of sugar alcohols. *Current opinion in biotechnology*, 19(5), pp.461–467. Available at: http://linkinghub.elsevier.com/retrieve/pii/S0958166908000943.

Alberice, J.V. et al., 2013. Searching for urine biomarkers of bladder cancer recurrence using a liquid chromatography-mass spectrometry and capillary electrophoresis-mass spectrometry metabolomics approach. *Journal of Chromatography A*, 1318, pp.163–170. Available at: http://dx.doi.org/10.1016/j.chroma.2013.10.002.

Andersen, M. & Kiel, P., 2000. Integrated utilisation of green biomass in the green biorefinery. *Industrial Crops and Products*, 11(2-3), pp.129–137.

Atkinson, D.E., 1968. The energy charge of the adenylate pool as a regulatory parameter. Interaction with feedback modifiers. *Biochemistry*, 7(11), pp.4030–4034.

Atkinson, D.E. & Walton, G.M., 1967. Adenosine triphosphate conservation in metabolic regulation. Rat liver citrate cleavage enzyme. *Journal of Biological Chemistry*, 242(13), pp.3239–3241.

Atkinson, S. et al., 1999. A hierarchical quorum-sensing system in *Yersinia pseudotuberculosis* is involved in the regulation of motility and clumping. *Molecular Microbiology*, 33(6), pp.1267–1277.

Balcke, G.U. et al., 2011. Linking energy metabolism to dysfunctions in mitochondrial respiration - A metabolomics in vitro approach. *Toxicology Letters*, 203(3), pp.200–209. Available at: http://dx.doi.org/10.1016/j.toxlet.2011.03.013.

Banerjee, S. et al., 2005. Comparison of *Mycobacterium tuberculosis* isocitrate dehydrogenases (ICD-1 and ICD-2) reveals differences in coenzyme affinity, oligomeric state, pH tolerance and phylogenetic affiliation. *BMC biochemistry*, 6(1), p.20. Available at: http://www.ncbi.nlm.nih.gov/pubmed/16194279.

Bar-Even, A. et al., 2010. Design and analysis of synthetic carbon fixation pathways. *Proceedings of the National Academy of Sciences of the United States of America*, 107(19), pp.8889–8894. Available at: http://www.pnas.org/cgi/doi/10.1073/pnas.0907176107.

Barnett, J.A., 2003. A history of research on yeasts 5: the fermentation pathway. *Yeast (Chichester, England)*, 20(6), pp.509–543. Available at: http://doi.wiley.com/10.1002/yea.986.

Bartsch, A., 2015. Metabolic network analysis of the marine bacterium *Dinoroseobacter shibae*. In C. Wittmann, ed. Available at: https://publikationen.sulb.uni-saarland.de/handle/20.500.11880/23197.

de Bary, A., 1884. *Vergleichende Morphologie und Biologie der Pilze, Mycetozoen und Bakterien.*, Leipzig: Wilhelm Engelmann.

Beales, N., 2004. Adaptation of microorganisms to cold temperatures, weak acid preservatives, low pH, and osmotic stress: A review. *Comprehensive Reviews in Food Science and Food Safety*, 3(1), pp.1–20. Available at: http://doi.wiley.com/10.1111/j.1541-4337.2004.tb00057.x.

Becker, J. et al., 2005. Amplified expression of fructose 1,6-bisphosphatase in *Corynebacterium glutamicum* increases in vivo flux through the pentose phosphate pathway and lysine production on different carbon sources. *Applied and Environmental Microbiology*, 71(12), pp.8587–8596.

Becker, J. et al., 2011. From zero to hero - design-based systems metabolic engineering of *Corynebacterium glutamicum* for L-lysine production. *Metabolic engineering*, 13(2), pp.159–168. Available at: http://linkinghub.elsevier.com/retrieve/pii/S1096717611000048.

Becker, J. et al., 2007. Metabolic flux engineering of L-lysine production in *Corynebacterium glutamicum* - over expression and modification of G6P dehydrogenase. *Journal of Biotechnology*, 132(2), pp.99–109. Available at: http://linkinghub.elsevier.com/retrieve/pii/S0168165607003860.

Becker, J. et al., 2010. Systems level engineering of *Corynebacterium glutamicum* - Reprogramming translational efficiency for superior production. *Engineering in Life Sciences*, 10(5), pp.430–438. Available at: http://onlinelibrary.wiley.com/doi/10.1002/elsc.201000008/abstract\nhttp://onlinelibrary.wiley.com/store/10.1002/elsc.201000008/asset/430_ftp.pdf?v=1&t=hy9pe0rr&s=81455dd70d1f6492ea6f82ec1882c145c0740308.

Becker, J. & Wittmann, C., 2015. Advanced biotechnology: metabolically engineered cells for the bio-based production of chemicals and fuels, materials, and health-care products. *Angewandte Chemie (International ed. in English)*, 54(11), pp.3328–3350. Available at: http://doi.wiley.com/10.1002/anie.201409033.

Becker, J. & Wittmann, C., 2012a. Bio-based production of chemicals, materials and fuels - *Corynebacterium glutamicum* as versatile cell factory. *Current opinion in biotechnology*, 23(4), pp.631–640. Available at: http://linkinghub.elsevier.com/retrieve/pii/S0958166911007233.

Becker, J. & Wittmann, C., 2012b. Systems and synthetic metabolic engineering for amino acid production - the heartbeat of industrial strain development. *Current opinion in biotechnology*, 23(5), pp.718–726. Available at: http://linkinghub.elsevier.com/retrieve/pii/S0958166911007683.

Bennett, B., Kimball, E. & Gao, M., 2009. Absolute metabolite concentrations and implied enzyme active site occupancy in *Escherichia coli*. *Nature chemical {...}*, 5(8), pp.593–599. Available at: http://www.nature.com/nchembio/journal/vaop/ncurrent/full/nchembio.186.html.

Benzer, S., 1961. On the topography of the genetic fine structure. *Proceedings of the National Academy of Sciences of the United States of America*, 47(3), pp.403–415.

Berger, A. et al., 2014. Robustness and plasticity of metabolic pathway flux among uropathogenic isolates of *Pseudomonas aeruginosa*. *PloS one*, 9(4), p.e88368. Available at: http://www.pubmedcentral.nih.gov/articlerender.fcgi?artid=3977821{&}tool=pmcentrez{&}rendertype=abstract.

Bessette, P.H. et al., 1999. Efficient folding of proteins with multiple disulfide bonds in the *Escherichia coli* cytoplasm. *Proceedings of the National Academy of Sciences of the United States of America*, 96(24), pp.13703–13708.

Bettenbrock, K. et al., 2007. Correlation between growth rates, EIIA^{Crr} phosphorylation, and intracellular cyclic AMP levels in *Escherichia coli* K-12. *Journal of Bacteriology*, 189(19), pp.6891–6900.

Biebl, H. et al., 2005. *Dinoroseobacter shibae* gen. nov., sp. nov., a new aerobic phototrophic bacterium isolated from dinoflagellates. *International Journal of Systematic and Evolutionary Microbiology*, 55(3), pp.1089–1096.

Blank, L.M. et al., 2008. Metabolic response of *Pseudomonas putida* during redox biocatalysis in the presence of a second octanol phase. *FEBS Journal*, 275(20), pp.5173–5190.

Blattner, F.R. et al., 1997. The complete genome sequence of *Escherichia coli* K-12. *Science (New York, N.Y.)*, 277(5331), pp.1453–1462.

Bligh, E.G. & Dyer, W.J., 1959. A rapid method of total lipid extraction and purification. *Canadian Journal of Biochemistry and Physiology*, 37(8), pp.911–917. Available at: http://www.nrcresearchpress.com/doi/abs/10.1139/o59-099.

Bolten, C.J., 2010. Bio-based Production of L-Methionine in *Corynebacterium glutamicum*. In C. Wittmann, ed. *Ibvt-Schriftenreihe, Band 48*. Cuvillier-Verlag.

Bolten, C.J. et al., 2007. Sampling for metabolome analysis of microorganisms. *Analytical Chemistry*, 79(10), pp.3843–3849.

Bolten, C.J. & Wittmann, C., 2008. Appropriate sampling for intracellular amino acid analysis in five phylogenetically different yeasts. *Biotechnology Letters*, 30(11), pp.1993–2000. Available at: http://link.springer.com/10.1007/s10529-008-9789-z.

Boynton, Z.L., Bennett, G.N. & Rudolph, F.B., 1994. Intracellular concentrations of coenzyme A and its derivatives from *Clostridium acetobutylicum* ATCC 824 and their roles in enzyme regulation. *Applied and Environmental Microbiology*, 60(1), pp.39–44.

Breitling, R., Pitt, A.R. & Barrett, M.P., 2006. Precision mapping of the metabolome. *Trends in biotechnology*, 24(12), pp.543–548. Available at: http://linkinghub.elsevier.com/retrieve/pii/S0167779906002642.

Broecker, S., Herre, S. & Pragst, F., 2012. General unknown screening in hair by liquid chromatography-hybrid quadrupole time-of-flight mass spectrometry (LC-QTOF-MS). *Forensic science international*, 218(1-3), pp.68–81. Available at: http://www.sciencedirect.com/science/article/pii/S0379073811004798.

Brubaker, R.R., 1968. Metabolism of carbohydrates by *Pasteurella pseudotuberculosis*. *Journal of Bacteriology*, 95(5), pp.1698–1705.

Buchholz, A. et al., 2002. Metabolomics: Quantification of intracellular metabolite dynamics. *Biomolecular Engineering*, 19(1), pp.5–15.

Buchinger, S. et al., 2009. A combination of metabolome and transcriptome analyses reveals new targets of the *Corynebacterium glutamicum* nitrogen regulator AmtR. *Journal of Biotechnology*, 140(1-2), pp.68–74.

Bücker, R., 2014. A multi-omics view on the pathogen Yersinia pseudotuberculosis - bridging metabolism and virulence. *C* Available at: http://scidok.sulb.uni-saarland.de/volltexte/2015/6239/.

Bücker, R. et al., 2014. The pyruvate-tricarboxylic acid cycle node: a focal point of virulence control in the enteric pathogen *Yersinia pseudotuberculosis*. *The Journal of biological chemistry*, 289(43), pp.30114–30132. Available at: http://www.jbc.org/lookup/doi/10.1074/jbc.M114.581348.

Bunk, B. et al., 2010. A short story about a big magic bug. *Bioengineered Bugs*, 1(2), pp.85–91.

Büscher, J.M. et al., 2009. Cross-platform comparison of methods for quantitative metabolomics of primary metabolism. *Analytical chemistry*, 81(6), pp.2135–2143. Available at: http://www.ncbi.nlm.nih.gov/pubmed/19236023.

Callahan, D.L. et al., 2009. Profiling of polar metabolites in biological extracts using diamond hydride-based aqueous normal phase chromatography. *Journal of Separation Science*, 32(13), pp.2273–2280.

Canelas, A.B. et al., 2009. Quantitative evaluation of intracellular metabolite extraction techniques for yeast metabolomics. *Analytical Chemistry*, 81(17), pp.7379–7389.

Del Castillo, T. et al., 2007. Convergent peripheral pathways catalyze initial glucose catabolism in *Pseudomonas putida*: Genomic and flux analysis. *Journal of Bacteriology*, 189(14), pp.5142–5152.

Castrillo, J.I. et al., 2003. An optimized protocol for metabolome analysis in yeast using direct infusion electrospray mass spectrometry. *Phytochemistry*, 62(6), pp.929–937.

Chain, P.S.G. et al., 2004. Insights into the evolution of *Yersinia pestis* through whole-genome comparison with *Yersinia pseudotuberculosis*. *Proceedings of the National Academy of Sciences of the United States of America*, 101(38), pp.13826–13831.

Chapman, A.G., Fall, L. & Atkinson, D.E., 1971. Adenylate energy charge in *Escherichia coli* during growth and starvation. *Journal of Bacteriology*, 108(3), pp.1072–1086.

Chohnan, S. et al., 1997. Changes in the size and composition of intracellular pools of nonesterified coenzyme A and coenzyme A thioesters in aerobic and facultatively anaerobic bacteria. *Applied and Environmental Microbiology*, 63(2), pp.553–560.

Christensen, B., Thykaer, J. & Nielsen, J., 2000. Metabolic characterization of high- and low-yielding strains of *Penicillium chrysogenum*. *Applied microbiology and biotechnology*, 54(2), pp.212–217. Available at: http://www.ncbi.nlm.nih.gov/pubmed/10968635.

Christie, G., Götzke, H. & Lowe, C.R., 2010. Identification of a receptor subunit and putative ligand-binding residues involved in the *Bacillus megaterium* QM B1551 spore germination response to glucose. *Journal of Bacteriology*, 192(17), pp.4317–4326.

Cirillo, V.P., 1961. Sugar transport in microorganisms. *Annual Review of Microbiology*, 15(1), pp.197–218. Available at: http://www.annualreviews.org/doi/abs/10.1146/annurev.mi.15.100161.001213.

Cornish-Bowden, A., 1981. Thermodynamic aspects of glycolysis. *Biochemistry and Molecular Biology*, 9(4), pp.133–137.

Cortassa, S., 1990. Thermodynamic and kinetic studies of a stoichiometric model of energetic metabolism under starvation conditions. *FEMS Microbiology Letters*, 66(1-3), pp.249–255. Available at: http://doi.wiley.com/10.1016/0378-1097(90)90292-X.

van Dam, J., 2002. Analysis of glycolytic intermediates in *Saccharomyces cerevisiae* using anion exchange chromatography and electrospray ionization with tandem mass spectrometric detection. *Analytica Chimica Acta*, 460(2), pp.209–218. Available at: http://linkinghub.elsevier.com/retrieve/pii/S0003267002002404.

Dauner, M. & Sauer, U., 2000. GC-MS analysis of amino acids rapidly provides rich information for isotopomer balancing. *Biotechnology Progress*, 16(4), pp.642–649.

Dauner, M., Storni, T. & Sauer, U., 2001. *Bacillus subtilis* metabolism and energetics in carbon-limited and excess-carbon chemostat culture. *Journal of Bacteriology*, 183(24), pp.7308–7317.

Dawes, E. a, 1988. Polyhydroxybutyrate: An intriguing biopolymer. *Bioscience Reports*, 8(6), pp.537–547. Available at: http://bioscirep.org/cgi/doi/10.1007/BF01117332.

Desagher, S., Glowinski, J. & Prémont, J., 1997. Pyruvate protects neurons against hydrogen peroxide-induced toxicity. *The Journal of neuroscience : the official journal of the Society for Neuroscience*, 17(23), pp.9060–9067. Available at: http://www.ncbi.nlm.nih.gov/pubmed/9364052.

DeTata, D., Collins, P. & McKinley, A., 2013. A fast liquid chromatography quadrupole time-of-flight mass spectrometry (LC-QToF-MS) method for the identification of organic explosives and propellants. *Forensic Science International*, 233(1-3), pp.63–74. Available at: http://dx.doi.org/10.1016/j.forsciint.2013.08.007.

Dietzler, D.N. et al., 1979a. Evidence for new factors in the coordinate regulation of energy metabolism in *Escherichia coli*. *The Journal of biological chemistry*, 254(17), pp.8295–8307. Available at: http://www.ncbi.nlm.nih.gov/pubmed/381303.

Dietzler, D.N. et al., 1979b. Evidence for new factors in the coordinate regulation of energy metabolism in *Escherichia coli*. Effects of hypoxia, chloramphenicol succinate, and 2,4-dinitrophenol on glucose utilization, glycogen synthesis, adenylate energy charge, and hexose pho. *Journal of Biological Chemistry*, 254(17), pp.8295–8307.

Dills, R.L. & Klaassen, C.D., 1986. The effect of inhibitors of mitochondrial energy production on hepatic glutathione, UDP-glucuronic acid, and adenosine 3'-phosphate-5'-phosphosulfate concentrations. *Drug metabolism and disposition: the biological fate of chemicals*, 14(2), pp.190–196. Available at: http://www.ncbi.nlm.nih.gov/pubmed/2870893.

Ducat, D.C. & Silver, P.A., 2012. Improving carbon fixation pathways. *Current opinion in chemical biology*, 16(3-4), pp.337–344. Available at: http://linkinghub.elsevier.com/retrieve/pii/S1367593112000634.

Dunn, W.B. & Ellis, D.I., 2005. Metabolomics: Current analytical platforms and methodologies. *TrAC Trends in Analytical Chemistry*, 24(4), pp.285–294. Available at: http://linkinghub.elsevier.com/retrieve/pii/S0165993605000348.

Dupler, M. & Baker, R., 1984. Survival of *Pseudomonas putida*, a biological control agent, in soil. *Phytopathology*, 74(2), pp.195–200.

Ebenhöh, O. & Heinrich, R., 2001. Evolutionary optimization of metabolic pathways. Theoretical reconstruction of the stoichiometry of ATP and NADH producing systems. *Bulletin of mathematical biology*, 63(1), pp.21–55.

Ebert, B.E. et al., 2011. Response of Pseudomonas putida KT2440 to increased NADH and ATP demand. *Applied and Environmental Microbiology*, 77(18), pp.6597–6605.

Ebert, M. et al., 2013. Transposon mutagenesis identified chromosomal and plasmid genes essential for adaptation of the marine bacterium *Dinoroseobacter shibae* to anaerobic conditions. *Journal of Bacteriology*, 195(20), pp.4769–4777.

Eley, J.H., Knobloch, K. & Han, T.W., 1979. Effect of growth condition on enzymes of the citric acid cycle and the glyoxylate cycle in the photosynthetic bacterium *Rhodopseudomonas palustris*. *Antonie van Leeuwenhoek*, 45(4), pp.521–529.

Empadinhas, N. & Da Costa, M.S., 2008. Osmoadaptation mechanisms in prokaryotes: Distribution of compatible solutes. *International Microbiology*, 11(3), pp.151–161.

Entner, N. & Doudoroff, M., 1952. Glucose and gluconic acid oxidation of *Pseudomonas saccharophila*. *The Journal of biological chemistry*, 196(2), pp.853–862. Available at: http://www.ncbi.nlm.nih.gov/pubmed/12981024.

Eppinger, M. et al., 2011. Genome sequences of the biotechnologically important *Bacillus megaterium* strains QM B1551 and DSM319. *Journal of Bacteriology*, 193(16), pp.4199–4213.

Far, J. et al., 2014. The Use of Ion Mobility Mass Spectrometry for Isomer Composition Determination Extracted from Se-Rich Yeast. *Analytical Chemistry*, 86, pp.11246–11254.

Feist, A.M. et al., 2007. A genome-scale metabolic reconstruction for *Escherichia coli* K-12 MG1655 that accounts for 1260 ORFs and thermodynamic information. *Molecular systems biology*, 3(121), pp.1–18. Available at: http://msb.embopress.org/cgi/doi/10.1038/msb4100155.

Fißler, J., Kohring, G.W. & Giffhorn, F., 1995. Enhanced hydrogen production from aromatic acids by immobilized cells of *Rhodopseudomonas palustris*. *Applied Microbiology and Biotechnology*, 44(1-2), pp.43–46.

Flamholz, A. et al., 2012. eQuilibrator - the biochemical thermodynamics calculator. *Nucleic acids research*, 40(Database issue), pp.D770–D775. Available at: http://nar.oxfordjournals.org/lookup/doi/10.1093/nar/gkr874.

Fleming, R.M.T., Thiele, I. & Nasheuer, H.P., 2009. Quantitative assignment of reaction directionality in constraint-based models of metabolism: application to *Escherichia coli*. *Biophysical chemistry*, 145(2-3), pp.47–56. Available at: http://linkinghub.elsevier.com/retrieve/pii/S0301462209001720.

Fliege, R. et al., 1992. The Entner-Doudoroff pathway in *Escherichia coli* is induced for oxidative glucose metabolism via pyrroloquinoline quinone-dependent glucose dehydrogenase. *Applied and Environmental Microbiology*, 58(12), pp.3826–3829.

Foerster, H.F. & Foster, J.W., 1966. Response of *Bacillus* spores to combinations of germinative compounds. *Journal of Bacteriology*, 91(3), pp.1168–1177.

Fuhrer, T., Fischer, E. & Sauer, U., 2005. Experimental identification and quantification of glucose metabolism in seven bacterial species. *Society*, 187(5), pp.1581–1590.

Fürch, T. et al., 2009. Metabolic fluxes in the central carbon metabolism of *Dinoroseobacter* shibae and *Phaeobacter gallaeciensis*, two members of the marine *Roseobacter* clade. *BMC microbiology*, 9, p.209.

Gabor, E. et al., 2011. The phosphoenolpyruvate-dependent glucose-phosphotransferase system from *Escherichia coli* K-12 as the center of a network regulating carbohydrate flux in the cell. *European Journal of Cell Biology*, 90(9), pp.711–720. Available at: http://dx.doi.org/10.1016/j.ejcb.2011.04.002.

Ganzera, M. et al., 2006. Determination of adenine and pyridine nucleotides in glucose-limited chemostat cultures of *Penicillium simplicissimum* by one-step ethanol extraction and ion-pairing liquid chromatography. *Analytical Biochemistry*, 359(1), pp.132–140.

Garg, S., Yang, L. & Mahadevan, R., 2010. Thermodynamic analysis of regulation in metabolic networks using constraint-based modeling. *BMC research notes*, 3(125), pp.1–7. Available at: http://www.pubmedcentral.nih.gov/articlerender.fcgi?artid=2873351&tool=pmcentrez&rendertype=abstract.

Gibbs, J.W., 1873. A method of geometrical representation of the thermodynamic properties of substances by means of surfaces. In *International Journal of Solids and Structures*. The Academy, pp. 382–404. Available at: http://pubs.acs.org/doi/abs/10.1021/ed300230v.

Godard, T., 2015. Systems biology of stress in *Bacillus megaterium* and its potential applications. In R. Krull, ed. *Ibvt-Schriftenreihe, Band 76*. Cuvillier-Verlag.

Göhler, A.-K. et al., 2011. More than just a metabolic regulator - elucidation and validation of new targets of PdhR in *Escherichia coli*. *BMC Systems Biology*, 5(1), p.197. Available at: http://www.biomedcentral.com/1752-0509/5/197.

Gomes, N.C.M. et al., 2005. Effects of the inoculant strain *Pseudomonas putida* KT2442 (pNF142) and of naphthalene contamination on the soil bacterial community. *FEMS Microbiology Ecology*, 54(1), pp.21–33.

Gonzalez, B., François, J. & Renaud, M., 1997. A rapid and reliable method for metabolite extraction in yeast using boiling buffered ethanol. *Yeast*, 13(14), pp.1347–1356.

Gravel, V., Antoun, H. & Tweddell, R.J., 2007. Growth stimulation and fruit yield improvement of greenhouse tomato plants by inoculation with *Pseudomonas putida* or *Trichoderma atroviride*: Possible role of indole acetic acid (IAA). *Soil Biology and Biochemistry*, 39(8), pp.1968–1977.

Greibrokk, T. & Andersen, T., 2003. High-temperature liquid chromatography. *Journal of Chromatography A*, 1000(1-2), pp.743–755. Available at: http://linkinghub.elsevier.com/retrieve/pii/S0021967302019635.

Guillarme, D. et al., 2010. Coupling ultra-high-pressure liquid chromatography with mass spectrometry. *TrAC Trends in Analytical Chemistry*, 29(1), pp.15–27. Available at: http://dx.doi.org/10.1016/j.trac.2009.09.008.

van Gulik, W.M., 2010. Fast sampling for quantitative microbial metabolomics. *Current Opinion in Biotechnology*, 21(1), pp.27–34. Available at: http://dx.doi.org/10.1016/j.copbio.2010.01.008.

Gunsalus, I.C., Horecker, B.L. & Wood, W.A., 1955. Pathways of carbohydrate metabolism in microorganisms. *Bacteriological reviews*, 19(2), pp.79–128. Available at: http://www.ncbi.nlm.nih.gov/pubmed/13239549.

Gustavsson, S.Å. et al., 2001. Studies of signal suppression in liquid chromatography-electrospray ionization mass spectrometry using volatile ion-pairing reagents. *Journal of Chromatography A*, 937(1-2), pp.41–47.

Hädicke, O., Grammel, H. & Klamt, S., 2011. Metabolic network modeling of redox balancing and biohydrogen production in purple nonsulfur bacteria. *BMC Systems Biology*, 5(1), p.150. Available at: http://www.biomedcentral.com/1752-0509/5/150.

Harrison, D.E., 1976. The regulation of respiration rate in growing bacteria. *Adv Microb Physiol*, 14(11), pp.243–313.

Hasunuma, T. et al., 2011. Metabolic pathway engineering based on metabolomics confers acetic and formic acid tolerance to a recombinant xylose-fermenting strain of *Saccharomyces cerevisiae*. *Microbial cell factories*, 10(2), pp.1–13. Available at: http://www.microbialcellfactories.com/content/10/1/2.

Haverkorn van Rijsewijk, B.R.B. et al., 2011. Large-scale ^{13}C-flux analysis reveals distinct transcriptional control of respiratory and fermentative metabolism in *Escherichia coli*. *Molecular systems biology*, 7(477), p.477. Available at: http://dx.doi.org/10.1038/msb.2011.9.

Hayes, T.L., Kenny, D. V & Hernon-Kenny, L., 2004. Feasibility of direct analysis of saliva and urine for phosphonic acids and thiodiglycol-related species associated with exposure to chemical warfare agents using LC-MS-MS. *J Med Chem Def*, 2(January), pp.1–23.

Hellerstein, M.K. et al., 1991. Measurement of de novo hepatic lipogenesis in humans using stable isotopes. *Journal of Clinical Investigation*, 87(5), pp.1841–1852. Available at: http://www.jci.org/articles/view/115206.

Henry, C.S., Broadbelt, L.J. & Hatzimanikatis, V., 2007. Thermodynamics-based metabolic flux analysis. *Biophysical journal*, 92(5), pp.1792–1805. Available at: http://linkinghub.elsevier.com/retrieve/pii/S0006349507709876.

Hess, B. & Boiteux, A., 1968. Mechanism of glycolytic oscillation in yeast. I. Aerobic and anaerobic growth conditions for obtaining glycolytic oscillation. *Hoppe-Seyler's Zeitschrift fur physiologische Chemie*, 349(11), pp.1567–1574.

Heumann, K.G., 1992. Isotope dilution mass spectrometry. *International Journal of Mass Spectrometry and Ion Processes*, 118-119(August 1991), pp.575–592.

Hiller, J., 2006. *Metabolische Analyse des Zentralstoffwechsels von* Escherichia coli., Available at: http://mediatum.ub.tum.de/doc/602000/document.pdf.

Hillier, S. & Charnetzky, W.T., 1981. Glyoxylate bypass enzymes in *Yersinia* species and multiple forms of isocitrate lyase in *Yersinia pestis*. *Journal of Bacteriology*, 145(1), pp.452–458.

Ho, C.S. et al., 2003. Electrospray ionisation mass spectrometry: principles and clinical applications. *The Clinical biochemist. Reviews / Australian Association of Clinical Biochemists*, 24(1), pp.3–12. Available at: http://www.ncbi.nlm.nih.gov/pubmed/18568044.

de Hoffmann, E., 1996. Tandem mass spectrometry: A primer. *Journal of Mass Spectrometry*, 31(2), pp.129–137. Available at: http://doi.wiley.com/10.1002/(SICI)1096-9888(199602)31:2<129::AID-JMS305>3.0.CO;2-T.

Hofmeyr, J.H.S. & Cornish-Bowden, A., 2000. Regulating the cellular economy of supply and demand. *FEBS Letters*, 476(1-2), pp.47–51.

Hogema, B.M. et al., 1998. Inducer exclusion in *Escherichia coli* by non-PTS substrates: The role of the PEP to pyruvate ratio in determining the phosphorylation state of enzyme IIA^{Glc}. *Molecular Microbiology*, 30(3), pp.487–498.

Holčapek, M. et al., 2004. Effects of ion-pairing reagents on the electrospray signal suppression of sulphonated dyes and intermediates. *Journal of Mass Spectrometry*, 39(1), pp.43–50.

Hoque, M. et al., 2011. Comparison of dynamic responses of cellular metabolites in *Escherichia coli* to pulse addition of substrates. *Biologia*, 66(6), pp.954–966. Available at: http://www.springerlink.com/index/10.2478/s11756-011-0136-9.

Hosono, K., 1992. Effect of salt stress on lipid composition and membrane fluidity of the salttolerant yeast *Zygosaccharomyces rouxii*. *Journal of General Microbiology*, 138(1), pp.91–96.

Huang, C.-J., Lin, H. & Yang, X., 2012. Industrial production of recombinant therapeutics in *Escherichia coli* and its recent advancements. *Journal of Industrial Microbiology & Biotechnology*, 39(3), pp.383–399.

Huck, J.H.J. et al., 2003. Profiling of pentose phosphate pathway intermediates in blood spots by tandem mass spectrometry: application to transaldolase deficiency. *Clinical chemistry*, 49(8), pp.1375–1380. Available at: http://www.ncbi.nlm.nih.gov/pubmed/12881455.

Imai, Y., Morita, S. & Arata, Y., 1984. Proton correlation NMR studies of metabolism in *Rhodopseudomonas palustris*. *Journal of biochemistry*, 96(3), pp.691–699.

Isaacson, M., Taylor, P. & Arntzen, L., 1983. Ecology of plague in Africa: Response of indigenous wild rodents to experimental plague infection. *Bulletin of the World Health Organization*, 61(2), pp.339–344.

Jackowski, S. & Rock, C.O., 1986. Consequences of reduced intracellular coenzyme A content in *Escherichia coli*. *Journal of Bacteriology*, 166(3), pp.866–871.

Jalava, K. et al., 2006. An outbreak of gastrointestinal illness and erythema nodosum from grated carrots contaminated with *Yersinia pseudotuberculosis*. *The Journal of infectious diseases*, 194(9), pp.1209–1216.

Jenkins, R.O., Stephens, G.M. & Dalton, H., 1987. Production of toluene *cis*-glycol by *Pseudomonas putida* in glucose feb-batch culture. *Biotechnology and Bioengineering*, 29(7), pp.873–883. Available at: http://www.ncbi.nlm.nih.gov/pubmed/18576532.

de Jonge, L. et al., 2014. Flux response of glycolysis and storage metabolism during rapid feast/famine conditions in *Penicillium chrysogenum* using dynamic ^{13}C labeling.

Biotechnology journal, 9(3), pp.372–385. Available at: http://www.ncbi.nlm.nih.gov/pubmed/24376125.

de Jonge, L.P. et al., 2012. Optimization of cold methanol quenching for quantitative metabolomics of *Penicillium chrysogenum*. *Metabolomics*, 8(4), pp.727–735. Available at: http://link.springer.com/10.1007/s11306-011-0367-3.

Jordan, K.W. et al., 2009. Metabolomic characterization of human rectal adenocarcinoma with intact tissue magnetic resonance spectroscopy. *Diseases of the colon and rectum*, 52(3), pp.520–5. Available at: http://www.ncbi.nlm.nih.gov/pubmed/19333056.

Joshi, G.S. et al., 2009. Differential accumulation of form I RubisCO in *Rhodopseudomonas palustris* CGA010 under photoheterotrophic growth conditions with reduced carbon sources. *Journal of Bacteriology*, 191(13), pp.4243–4250.

Kapatral, V., Bina, X. & Chakrabarty, a M., 2000. Succinyl coenzyme A synthetase of *Pseudomonas aeruginosa* with a broad specificity for nucleoside triphosphate (NTP) synthesis modulates specificity for NTP synthesis by the 12-kilodalton form of nucleoside diphosphate kinase. *Journal of bacteriology*, 182(5), pp.1333–1339. Available at: http://www.pubmedcentral.nih.gov/articlerender.fcgi?artid=94420{&}tool=pmcentrez{&}rendertype=abstract.

Kelleher, J.K. et al., 1994. Isotopomer spectral analysis of cholesterol synthesis: applications in human hepatoma cells. *The American journal of physiology*, 266(3 Pt 1), pp.384–395. Available at: http://www.ncbi.nlm.nih.gov/pubmed/8166258.

Kelly, G.J. & Turner, J.F., 1969. The regulation of pea-seed phosphofructokinase by phosphoenolpyruvate. *The Biochemical journal*, 115(3), pp.481–487.

Khan, J., 2011. Biodegradation of azo dye by moderately halotolerant *Bacillus megaterium* and study of enzyme azoreductase involved in degradation. *Adv. Biotech*, 10(07), pp.21–27. Available at: http://www.advancedbiotech.in/Jan{_}11.

Kiefer, P. et al., 2007. Determination of carbon labeling distribution of intracellular metabolites from single fragment ions by ion chromatography tandem mass spectrometry. *Analytical biochemistry*, 360(2), pp.182–188. Available at: http://linkinghub.elsevier.com/retrieve/pii/S0003269706004684.

Kim, J.K. & Lee, B.K., 2000. Mass production of *Rhodopseudomonas palustris* as diet for aquaculture. *Aquacultural Engineering*, 23(4), pp.281–293.

Klemme, J.-H., Chyla, I. & Preuss, M., 1980. Dissimilatory nitrate reduction by strains of the facultative phototrophic bacterium *Rhodopseudomonas palustris*. *FEMS Microbiology Letters*, 9(2), pp.137–140. Available at: http://femsle.oxfordjournals.org/cgi/doi/10.1111/j.1574-6968.1980.tb05623.x.

Klingner, A. et al., 2015. Large-scale ^{13}C flux profiling reveals conservation of the Entner-Doudoroff pathway as a glycolytic strategy among marine bacteria that use glucose. H. Nojiri, ed. *Applied and Environmental Microbiology*, 81(7), pp.2408–2422. Available at: http://aem.asm.org/lookup/doi/10.1128/AEM.03157-14.

Klungsoyr, L. et al., 1968. Interaction between energy charge and product feedback in the regulation of biosynthetic enzymes. Aspartokinase, phosphoribosyladenosine triphosphate synthetase, and phosphoribosyl pyrophosphate synthetase. *Biochemistry*, 7(11), pp.4035–4040.

Kohlstedt, M., 2014. *A multi-omics perspective on osmoadaptation and osmoprotection in* Bacillus subtilis., *IsBio-Schriftenreihe, Band 1*. Cuvillier-Verlag. Available at: http://scidok.sulb.uni-saarland.de/volltexte/2014/5867/.

Kohlstedt, M. et al., 2014. Adaptation of *Bacillus subtilis* carbon core metabolism to simultaneous nutrient limitation and osmotic challenge: a multi-omics perspective. *Environmental microbiology*, 16(6), pp.1898–1917. Available at: http://www.ncbi.nlm.nih.gov/pubmed/24571712.

Kohlstedt, M., Becker, J. & Wittmann, C., 2010. Metabolic fluxes and beyond - systems biology understanding and engineering of microbial metabolism. *Applied Microbiology and Biotechnology*, 88(5), pp.1065–1075. Available at: http://link.springer.com/10.1007/s00253-010-2854-2.

de Koning, W. & van Dam, K., 1992. A method for the determination of changes of glycolytic metabolites in yeast on a subsecond time scale using extraction at neutral pH. *Analytical biochemistry*, 204(1), pp.118–123. Available at: http://www.ncbi.nlm.nih.gov/pubmed/1514678.

Kornberg, H.L., 1966. The role and control of the glyoxylate cycle in *Escherichia coli*. *The Biochemical journal*, 99(1), pp.1–11. Available at: http://www.ncbi.nlm.nih.gov/pubmed/5337756.

Korneli, C. et al., 2013. Getting the big beast to work - Systems biotechnology of *Bacillus megaterium* for novel high-value proteins. *Journal of Biotechnology*, 163(2), pp.87–96. Available at: http://dx.doi.org/10.1016/j.jbiotec.2012.06.018.

Kotrba, P., Inui, M. & Yukawa, H., 2001. Bacterial phosphotransferase system (PTS) in carbohydrate uptake and control of carbon metabolism. *Journal of Bioscience and Bioengineering*, 92(6), pp.502–517.

Krebs, H. a. & Johnson, W. a., 1980. The role of citric acid in intermediate metabolism in animal tissues. *FEBS Letters*, 117 Suppl(August), pp.148–156.

Krömer, J.O. et al., 2004. In-depth profiling of lysine-producing *Corynebacterium glutamicum* by combined analysis of the transcriptome, metabolome, and fluxome. *Journal of bacteriology*, 186(6), pp.1769–1784. Available at: http://www.ncbi.nlm.nih.gov/pubmed/14996808.

Kümmel, A., Panke, S. & Heinemann, M., 2006. Systematic assignment of thermodynamic constraints in metabolic network models. *BMC bioinformatics*, 7(512). Available at: http://www.pubmedcentral.nih.gov/articlerender.fcgi?artid=1664590&tool=pmcentrez&rendertype=abstract\nhttp://www.ncbi.nih.gov/entrez/query.fcgi?cmd=retrieve&db=pubmed&dopt=abstract&list_uids=17123434.

Kuze, Y. et al., 2013. Highly sensitive liquid chromatography-tandem mass spectrometry method for quantification of TAK-448 in human plasma. *Journal of pharmaceutical and biomedical analysis*, 83, pp.75–81. Available at: http://dx.doi.org/10.1016/j.jpba.2013.04.023.

Laass, S. et al., 2014. Gene regulatory and metabolic adaptation processes of *Dinoroseobacter shibae* DFL12^T during oxygen depletion. *Journal of Biological Chemistry*, 289(19), pp.13219–13231.

Lameiras, F., Heijnen, J.J. & van Gulik, W.M., 2015. Development of tools for quantitative intracellular metabolomics of *Aspergillus niger* chemostat cultures. *Metabolomics*, 11(5), pp.1253–1264. Available at: http://link.springer.com/10.1007/s11306-015-0781-z.

Lämmerhofer, M. & Weckwerth, W., 2013. Metabolomics in Practice: Succesful Strategies to Generate and Analyze Metabolic Data. In M. Lämmerhofer & W. Weckwerth, eds. Wiley-VCH Verlag GmbH & Co. KGaA.

Larimer, F.W. et al., 2004. Complete genome sequence of the metabolically versatile photosynthetic bacterium *Rhodopseudomonas palustris*. *Nature biotechnology*, 22(1), pp.55–61.

Lederberg, J. & Tatum, E.L., 1946. Gene recombination in *Escherichia coli*. *Nature*, 158(4016), p.558.

Lee, S.Y., Lee, D.-Y. & Kim, T.Y., 2005. Systems biotechnology for strain improvement. *Trends in biotechnology*, 23(7), pp.349–358. Available at: http://linkinghub.elsevier.com/retrieve/pii/S0167779905001150.

Leuchtenberger, W., Huthmacher, K. & Drauz, K., 2005. Biotechnological production of amino acids and derivatives: Current status and prospects. *Applied Microbiology and Biotechnology*, 69(1), pp.1–8.

Li, F., Gonzalez, F.J. & Ma, X., 2012. LC-MS-based metabolomics in profiling of drug metabolism and bioactivation. *Acta Pharmaceutica Sinica B*, 2(2), pp.116–123. Available at: http://dx.doi.org/10.1016/j.apsb.2012.02.010.

Li, X. et al., 2011. A database of thermodynamic properties of the reactions of glycolysis, the tricarboxylic acid cycle, and the pentose phosphate pathway. *Database*, 2011, pp.1–15.

Lippe, K., 2008. Medienoptimierung und Kultivierung des phototrophen Mikroorganismusses *Rhodopseudomonas palustris*. In E. Franco-Lara, ed.

Logsdon, L.K. & Mecsas, J., 2006. The proinflammatory response induced by wild-type *Yersinia pseudotuberculosis* infection inhibits survival of *yop* mutants in the gastrointestinal tract and Peyer's patches. *Infection and Immunity*, 74(3), pp.1516–1527.

Los, D. a. & Murata, N., 2004. Membrane fluidity and its roles in the perception of environmental signals. *Biochimica et Biophysica Acta - Biomembranes*, 1666(1-2), pp.142–157.

Lüders, S. et al., 2011. Influence of the hydromechanical stress and temperature on growth and antibody fragment production with *Bacillus megaterium*. *Applied Microbiology and Biotechnology*, 91(1), pp.81–90.

Luo, B. et al., 2007. Simultaneous determination of multiple intracellular metabolites in glycolysis, pentose phosphate pathway and tricarboxylic acid cycle by liquid chromatography-mass spectrometry. *Journal of chromatography. A*, 1147(2), pp.153–164. Available at: http://linkinghub.elsevier.com/retrieve/pii/S0021967307002907.

Lynn, K.-S. et al., 2015. Metabolite Identification for Mass Spectrometry-Based Metabolomics Using Multiple Types of Correlated Ion Information. *Analytical Chemistry*, 87(4), pp.2143–2151. Available at: <Go to ISI>://WOS:000349806200021\nhttp://pubs.acs.org/doi/pdfplus/10.1021/ac503325c.

Makarov, A. & Scigelova, M., 2010. Coupling liquid chromatography to Orbitrap mass spectrometry. *Journal of chromatography. A*, 1217(25), pp.3938–3945. Available at: http://linkinghub.elsevier.com/retrieve/pii/S002196731000227X.

Marcellin, E. et al., 2009. Quantitative analysis of intracellular sugar phosphates and sugar nucleotides in encapsulated streptococci using HPAEC-PAD. *Biotechnology Journal*, 4(1), pp.58–63.

Marriott, I.D., Dawes, E.A. & Rowley, B.I., 1981. Effect of growth rate and nutrient limitation on the adenine nucleotide content, energy charge and enzymes of adenylate metabolism in *Azotobacter beijerinckii*. *Microbiology*, 125(2), pp.375–382. Available at: http://mic.microbiologyresearch.org/content/journal/micro/10.1099/00221287-125-2-375.

Martínez, V.S. & Nielsen, L.K., 2014. NExT: integration of thermodynamic constraints and metabolomics data into a metabolic network. *Methods in molecular biology (Clifton, N.J.)*, 1191(1), pp.65–78. Available at: http://www.ncbi.nlm.nih.gov/pubmed/25178784.

Marx, A. et al., 1996. Determination of the fluxes in the central metabolism of *Corynebacterium glutamicum* by nuclear magnetic resonance spectroscopy combined with metabolite balancing. *Biotechnology and Bioengineering*, 49(2), pp.111–129.

Mashego, M.R. et al., 2007. Microbial metabolomics: Past, present and future methodologies. *Biotechnology Letters*, 29(1), pp.1–16.

Maurer, H.H., 2005. Multi-analyte procedures for screening for and quantification of drugs in blood, plasma, or serum by liquid chromatography-single stage or tandem mass spectrometry (LC-MS or LC-MS/MS) relevant to clinical and forensic toxicology. *Clinical Biochemistry*, 38(4), pp.310–318.

Mavrovouniotis, M.L., 1993. Identification of localized and distributed bottlenecks in metabolic pathways. *Proceedings / ... International Conference on Intelligent Systems for Molecular Biology ; ISMB. International Conference on Intelligent Systems for Molecular Biology*, 1, pp.275–283.

McKinlay, J.B. & Harwood, C.S., 2010. Carbon dioxide fixation as a central redox cofactor recycling mechanism in bacteria. *Proceedings of the National Academy of Sciences*, 107(26), pp.11669–11675. Available at: http://www.pnas.org/cgi/doi/10.1073/pnas.1006175107.

Meléndez-Hevia, E. et al., 1997. Theoretical approaches to the evolutionary optimization of glycolysis: Chemical analysis. *European journal of biochemistry / FEBS*, 244(2), pp.527–543. Available at: http://www.ncbi.nlm.nih.gov/pubmed/9119021.

Mercado-Lubo, R. et al., 2008. A *Salmonella enterica* serovar typhimurium succinate dehydrogenase/fumarate reductase double mutant is avirulent and immunogenic in BALB/c mice. *Infection and Immunity*, 76(3), pp.1128–1134.

Mercado-Lubo, R. et al., 2009. Salmonella enterica serovar typhimurium mutants unable to convert malate to pyruvate and oxaloacetate are avirulent and immunogenic in BALB/c mice. *Infection and Immunity*, 77(4), pp.1397–1405.

De Mey, M. et al., 2010. Catching prompt metabolite dynamics in *Escherichia coli* with the BioScope at oxygen rich conditions. *Metabolic Engineering*, 12(5), pp.477–487. Available at: http://dx.doi.org/10.1016/j.ymben.2010.04.003.

Molt, K. & Telgheder, U., 2010. *Berechnung der Verfahrensstandardabweichung und Bestimmungsgrenze aus einer Kalibrierung gem{ä}{ß} DIN 32645.*, Available at: https://www.google.de/url?sa=t{&}rct=j{&}q={&}esrc=s{&}source=web{&}cd=1{&}ved=0CCIQFjAA{&}url=https://www.uni-due.de/imperia/md/content/iac/git{_}erw{_}1.pdf{&}ei=XEyEVcDlD4u5ygOsxlTABw{&}usg=AFQjCNHIAVxO0VECQQJpO-J5cQgDzpKmiA{&}bvm=bv.96339352,d.bGQ{&}c.

Moritz, B. et al., 2002. Changes of pentose phosphate pathway flux in vivo in *Corynebacterium glutamicum* during leucine-limited batch cultivation as determined from intracellular metabolite concentration measurements. *Metabolic Engineering*, 4(4), pp.295–305. Available at: http://www.ncbi.nlm.nih.gov/pubmed/12646324.

Mortier, K.A. et al., 2004. Adduct formation in quantitative bioanalysis: effect of ionization conditions on paclitaxel. *Journal of the American Society for Mass Spectrometry*, 15(4), pp.585–592. Available at: http://link.springer.com/10.1016/j.jasms.2003.12.013.

Nedwell, D., 1999. Effect of low temperature on microbial growth: lowered affinity for substrates limits growth at low temperature. *FEMS microbiology ecology*, 30(2), pp.101–111. Available at: http://www.ncbi.nlm.nih.gov/pubmed/10508935.

Nelson, K.E. et al., 2002. Complete genome sequence and comparative analysis of the metabolically versatile *Pseudomonas putida* KT2440. *Environmental Microbiology*, 4(12), pp.799–808. Available at: <Go.

Nicholson, J.K., 2006. Global systems biology, personalized medicine and molecular epidemiology. *Molecular systems biology*, 2(52), pp.1–6.

Nikel, P.I. et al., 2015. *Pseudomonas putida* KT2440 strain metabolizes glucose through a cycle formed by enzymes of the Entner-Doudoroff, Embden-Meyerhof-Parnas, and pentose phosphate pathways. *Journal of Biological Chemistry*, 290(43), pp.25920–25932. Available at: http://www.jbc.org/lookup/doi/10.1074/jbc.M115.687749.

Noack, S. & Wiechert, W., 2014. Quantitative metabolomics: A phantom? *Trends in Biotechnology*, 32(5), pp.238–244. Available at: http://dx.doi.org/10.1016/j.tibtech.2014.03.006.

Noor, E. et al., 2010. Central carbon metabolism as a minimal biochemical walk between precursors for biomass and energy. *Molecular Cell*, 39(5), pp.809–820. Available at: http://dx.doi.org/10.1016/j.molcel.2010.08.031.

Oldiges, M. et al., 2004. Stimulation, monitoring, and analysis of pathway dynamics by metabolic profiling in the aromatic amino acid pathway. *Biotechnology Progress*, 20, pp.1623–1633.

Otani, M. et al., 1987. Gluconate metabolism in germinated spores of *Bacillus megaterium* QM B1551: primary roles of gluconokinase and the pentose cycle. *Biochimica et biophysica acta*, 924(3), pp.467–472.

Otani, M. et al., 1986. Predominance of gluconate formation from glucose during germination of *Bacillus megaterium* QM B1551 spores. *Journal of Bacteriology*, 167(1), pp.148–152.

Park, S.M. et al., 1997. Elucidation of anaplerotic pathways in *Corynebacterium glutamicum* via ^{13}C-NMR spectroscopy and GC-MS. *Applied Microbiology and Biotechnology*, 47(4), pp.430–440. Available at: http://dx.doi.org/10.1007/s002530050952.

Pedersen, Carlsen & Nielsen, 1999. Identification of enzymes and quantification of metabolic fluxes in the wild type and in a recombinant *Aspergillus oryzae* strain. *Applied and environmental microbiology*, 65(1), pp.11–19. Available at: http://www.ncbi.nlm.nih.gov/pubmed/9872753.

Perrenoud, A. & Sauer, U., 2005. Impact of global transcriptional regulation by ArcA, ArcB, Cra, Crp, Cya, Fnr, and Mlc on glucose catabolism in *Escherichia coli*. *Journal of Bacteriology*, 187(9), pp.3171–3179.

Pesek, J.J. & Matyska, M.T., 2005. Hydride-based silica stationary phases for HPLC: Fundamental properties and applications. *Journal of Separation Science*, 28(15), pp.1845–1854.

Pette, D. & Reichmann, H., 1982. A method for quantitative extraction of enzymes and metabolites from tissue samples in the milligram range. *The journal of histochemistry and cytochemistry : official journal of the Histochemistry Society*, 30(4), pp.401–402. Available at: http://www.ncbi.nlm.nih.gov/pubmed/7061832.

Pettit, F.H., Pelley, J.W. & Reed, L.J., 1975. Regulation of pyruvate dehydrogenase kinase and phosphatase by acetyl-CoA/CoA and NADH/NAD ratios. *Biochemical and biophysical research communications*, 65(2), pp.575–582. Available at: http://www.ncbi.nlm.nih.gov/pubmed/167775.

Peyraud, R. et al., 2009. Demonstration of the ethylmalonyl-CoA pathway by using ^{13}C metabolomics. *Proceedings of the National Academy of Sciences of the United States of America*, 106(12), pp.4846–4851.

Plumb, R.S. et al., 2005. A rapid screening approach to metabonomics using UPLC and oa-TOF mass spectrometry: application to age, gender and diurnal variation in normal/Zucker obese rats and black, white and nude mice. *The Analyst*, 130(6), pp.844–849.

Poblete Castro, I.A., 2012. Systems biotechnology of *Pseudomonas putida* for the enhanced production of polyhydroxyalkanoates: a rational approach for strain and bioprocess engineering. In C. Wittmann, ed. *Ibvt-Schriftenreihe*. Cuvillier-Verlag.

Poblete-Castro, I. et al., 2012. Industrial biotechnology of *Pseudomonas putida* and related species. *Applied Microbiology and Biotechnology*, 93(6), pp.2279–2290.

Polce, M.J. et al., 1996. Characterization of neutral fragments in tandem mass spectrometry: A unique route to mechanistic and structural information. *Journal of Mass Spectrometry*, 31(10), pp.1073–1085.

Qian, H. & Beard, D.A., 2006. Metabolic futile cycles and their functions: a systems analysis of energy and control. *Systems biology*, 153(4), pp.192–200.

Quinones, M.P. & Kaddurah-Daouk, R., 2009. Metabolomics tools for identifying biomarkers for neuropsychiatric diseases. *Neurobiology of Disease*, 35(2), pp.165–176. Available at: http://dx.doi.org/10.1016/j.nbd.2009.02.019.

Rabinowitz, J.D. & Kimball, E., 2007. Acidic acetonitrile for cellular metabolome extraction from *Escherichia coli*. *Analytical Chemistry*, 79(16), pp.6167–6173.

Ramaiah, A., Hathaway, J.A. & Atkinson, D.E., 1964. Adenylate as a metabolic regulator: Effect on yeast phosphofructokinase kinetics. *The Journal of biological chemistry*, 239, pp.3619–3622.

Ratledge, C., 2004. Fatty acid biosynthesis in microorganisms being used for Single Cell Oil production. *Biochimie*, 86, pp.807–815.

Reeve, C.A., Bockman, A.T. & Matin, A., 1984. Role of protein degradation in the survival of carbon-starved *Escherichia coli* and *Salmonella typhimurium*. *Journal of bacteriology*, 157(3), pp.758–763. Available at: http://www.pubmedcentral.nih.gov/articlerender.fcgi?artid=215323{&}tool=pmcentrez{&}rendertype=abstract.

Rex, R. et al., 2013. Swimming in light: a large-scale computational analysis of the metabolism of *Dinoroseobacter shibae*. *PLoS Computational Biology*, 9(10).

Rodrigues, A.L. et al., 2014. Systems metabolic engineering of *Escherichia coli* for gram scale production of the antitumor drug deoxyviolacein from glycerol. *Biotechnology and Bioengineering*, 111(11), pp.2280–2289. Available at: http://doi.wiley.com/10.1002/bit.25297.

Romano, A.H. & Conway, T., 1996. Evolution of carbohydrate metabolic pathways. *Research in Microbiology*, 147(6-7), pp.448–455.

Romualdi, C. & Gerolamo, L., 2009. Statistical tools for gene expression analysis and systems biology and related web resources. In S. Krawetz, ed. *Bioinformatics for Systems Biology*. Humana Press, pp. 181–205. Available at: http://link.springer.com/chapter/10.1007/978-1-59745-440-7{_}11{#}page-1.

Ronchel, M.C. et al., 1995. Construction and behavior of biologically contained bacteria for environmental applications in bioremediation. *Applied and Environmental Microbiology*, 61(8), pp.2990–2994.

Roszak, A.W. et al., 2003. Crystal structure of the RC-LH1 core complex from *Rhodopseudomonas palustris*. *Science (New York, N.Y.)*, 302(5652), pp.1969–1972.

Rubia, T. et al., 1986. Adenine nucleotide content and energy charge of *Bacillus megaterium* during batch growth in low-phosphate medium. *FEMS Microbiology Letters*, 35(1), pp.5–9. Available at: http://femsle.oxfordjournals.org/cgi/doi/10.1111/j.1574-6968.1988.tb02992.x.

Rui, B. et al., 2010. A systematic investigation of *Escherichia coli* central carbon metabolism in response to superoxide stress. *BMC systems biology*, 4, p.122.

Russo, E., 2003. Learning how to manipulate DNA's double helix has fuelled job growth in biotechnology during the past 50 years, says Eugene Russo. *Nature*, 421, pp.456–457.

Sauer, U., 2004. High-throughput phenomics: Experimental methods for mapping fluxomes. *Current Opinion in Biotechnology*, 15(1), pp.58–63.

Sauer, U. & Eikmanns, B.J., 2005. The PEP-pyruvate-oxaloacetate node as the switch point for carbon flux distribution in bacteria. *FEMS Microbiology Reviews*, 29(4), pp.765–794.

Schneider, D. a, Gaal, T. & Gourse, R.L., 2002. NTP-sensing by rRNA promoters in *Escherichia coli* is direct. *Proceedings of the National Academy of Sciences of the United States of America*, 99(13), pp.8602–8607.

Schug, K. & McNair, H.M., 2002. Adduct formation in electrospray ionization. Part 1: Common acidic pharmaceuticals. *Journal of Separation Science*, 25(12), pp.759–766. Available at: http://dx.doi.org/10.1002/1615-9314(20020801)25:12<759::AID-JSSC760>3.0.CO;2-M.

Setlow, B. & Setlow, P., 1977. Levels of oxidized and reduced pyridine nucleotides in dormant spores and during growth, sporulation, and spore germination of *Bacillus megaterium*. *Journal of bacteriology*, 129(2), pp.857–865. Available at: http://www.pubmedcentral.nih.gov/articlerender.fcgi?artid=235022{&}tool=pmcentrez{&}rendertype=abstract.

Shen, L.C. et al., 1968. Interaction between energy charge and metabolite modulation in the regulation of enzymes of amphibolic sequences. Phosphofructokinase and pyruvate dehydrogenase. *Biochemistry*, 7(11), pp.4041–4045.

Shulman, S.T., Friedmann, H.C. & Sims, R.H., 2007. Theodor Escherich: the first pediatric infectious diseases physician? *Clinical infectious diseases : an official publication of the Infectious Diseases Society of America*, 45(8), pp.1025–1029.

Smith, E. & Morowitz, H.J., 2004. Universality in intermediary metabolism. *Proceedings of the National Academy of Sciences of the United States of America*, 101(36), pp.13168–13173.

Snyder, L.R., Kirkland, J.J. & Dolan, J.W., 2010. *Introduction to modern liquid chromatography.*, Available at: http://link.springer.com/10.1007/s00216-010-4483-0.

Sonntag, D. et al., 2011. Targeted metabolomics for bioprocessing. *BMC proceedings*, 5 Suppl 8(Suppl 8), p.P27. Available at: http://www.pubmedcentral.nih.gov/articlerender.fcgi?artid=3284902{&}tool=pmcentrez{&}rendertype=abstract.

Soora, M. & Cypionka, H., 2013. Light enhances survival of *Dinoroseobacter shibae* during long-term starvation. *PLoS ONE*, 8(12), pp.6–12.

Spaggiari, D., Geiser, L. & Rudaz, S., 2014. Coupling ultra-high-pressure liquid chromatography with mass spectrometry for in-vitro drug-metabolism studies. *TrAC Trends in Analytical Chemistry*, 63, pp.129–139. Available at: http://linkinghub.elsevier.com/retrieve/pii/S0165993614001873.

Spagou, K. et al., 2011. HILIC-UPLC-MS for exploratory urinary metabolic profiling in toxicological studies. *Analytical Chemistry*, 83(1), pp.382–390.

Spiro, S. & Guest, J.R., 1991. Adaptive responses to oxygen limitation in *Escherichia coli*. *Trends in biochemical sciences*, 16(8), pp.310–314.

Swedes, J.S., Sedo, R.J. & Atkinson, D.E., 1975. Relation of growth and protein synthesis to the adenylate energy charge in an adenine-requiring mutant of *Escherichia coli*. *The*

Journal of biological chemistry, 250(17), pp.6930–6938. Available at: http://www.ncbi.nlm.nih.gov/pubmed/1099099.

Tännler, S. et al., 2008. CcpN controls central carbon fluxes in *Bacillus subtilis*. *Journal of bacteriology*, 190(18), pp.6178–6187. Available at: http://www.scopus.com/inward/record.url?eid=2-s2.0-51549110350{&}partnerID=tZOtx3y1.

Taymaz-Nikerel, H. et al., 2009. Development and application of a differential method for reliable metabolome analysis in *Escherichia coli*. *Analytical Biochemistry*, 386(1), pp.9–19. Available at: http://dx.doi.org/10.1016/j.ab.2008.11.018.

Theodoridis, G.A. et al., 2012. Liquid chromatography-mass spectrometry based global metabolite profiling: a review. *Analytica chimica acta*, 711, pp.7–16. Available at: http://linkinghub.elsevier.com/retrieve/pii/S0003267011013109.

Thevis, M., Thomas, A. & Schänzer, W., 2008. Mass spectrometric determination of insulins and their degradation products in sports drug testing. *Mass Spectrometry Reviews*, 27(1), pp.35–50.

Tomasch, J. et al., 2011. Transcriptional response of the photoheterotrophic marine bacterium *Dinoroseobacter shibae* to changing light regimes. *The ISME Journal*, 5(12), pp.1957–1968.

Tonouchi, N., Sugiyama, M. & Yokozeki, K., 2003. Coenzyme specificity of enzymes in the oxidative pentose phosphate pathway of *Gluconobacter oxydans*. *Bioscience, biotechnology, and biochemistry*, 67(12), pp.2648–2651. Available at: <Go.

Touchstone, J.C., 1993. History of Chromatography. *Journal of Liquid Chromatography*, 16(8), pp.1647–1665. Available at: http://www.tandfonline.com/doi/abs/10.1080/10826079308021679.

Toya, Y. & Shimizu, H., 2013. Flux analysis and metabolomics for systematic metabolic engineering of microorganisms. *Biotechnology Advances*, 31(6), pp.818–826. Available at: http://linkinghub.elsevier.com/retrieve/pii/S0734975013000839.

Trethewey, R.N., 2004. Metabolite profiling as an aid to metabolic engineering in plants. *Current Opinion in Plant Biology*, 7(2), pp.196–201.

Trotter, E.W. et al., 2011. Reprogramming of *Escherichia coli* K-12 metabolism during the initial phase of transition from an anaerobic to a micro-aerobic environment. *PLoS ONE*, 6(9), p.e25501.

Troxell, B. et al., 2014. Pyruvate protects pathogenic spirochetes from H_2O_2 killing. R. M. Wooten, ed. *PloS one*, 9(1), p.e84625. Available at: http://www.pubmedcentral.nih.gov/articlerender.fcgi?artid=3879313{&}tool=pmcentrez{&}rendertype=abstract.

Udaka, S., 1960. Screening method for microorganisms accumulating metabolites and its use in the isolation of *Micrococcus glutamicus*. *Journal of bacteriology*, 79, pp.754–755.

Vallino, J.J. & Stephanopoulos, G., 1994. Carbon flux distributions at the glucose 6-phosphate branch point in *Corynebacterium glutamicum* during lysine overproduction. *Biotechnology Progress*, 10(3), pp.327–334. Available at: http://onlinelibrary.wiley.com/doi/10.1021/bp00027a014/abstract.

Vary, P.S. et al., 2007. *Bacillus megaterium* - from simple soil bacterium to industrial protein production host. *Applied Microbiology and Biotechnology*, 76(5), pp.957–967.

Vemuri, G.N. et al., 2007. Increasing NADH oxidation reduces overflow metabolism in *Saccharomyces cerevisiae*. *Proceedings of the National Academy of Sciences*, 104(7), pp.2402–2407. Available at: http://www.ncbi.nlm.nih.gov/pubmed/17287356.

Vertes, A.A., Inui, M. & Yukawa, H., 2005. Manipulating corynebacteria, from individual genes to chromosomes. *Appl Environ Microbiol*, 71(12), pp.7633–7642. Available at: http://www.ncbi.nlm.nih.gov/entrez/query.fcgi?cmd=Retrieve&db=PubMed&dopt=Citation&list_uids=16332735\nhttp://pubmedcentralcanada.ca/picrender.cgi?accid=PMC1317429&blobtype=pdf.

Villas-Bôas, S.G. et al., 2005. Global metabolite analysis of yeast: Evaluation of sample preparation methods. *Yeast*, 22(14), pp.1155–1169.

Vojinović, V. & von Stockar, U., 2009. Influence of uncertainties in pH, pMg, activity coefficients, metabolite concentrations, and other factors on the analysis of the thermodynamic feasibility of metabolic pathways. *Biotechnology and bioengineering*, 103(4), pp.780–795. Available at: http://www.ncbi.nlm.nih.gov/pubmed/19365870.

Vu-Trong, K. & Gray, P.P., 1982. Continuous-culture studies on the regulation of tylosin biosynthesis. *Biotechnology and Bioengineering*, 24(5), pp.1093–1103. Available at: http://doi.wiley.com/10.1002/bit.260240506.

Wacker, M. et al., 2002. N-linked glycosylation in *Campylobacter jejuni* and its functional transfer into *E. coli*. *Science (New York, N.Y.)*, 298(5599), pp.1790–1793.

Wagner-Döbler, I. et al., 2010. The complete genome sequence of the algal symbiont *Dinoroseobacter shibae*: a hitchhiker's guide to life in the sea. *The ISME journal*, 4(1), pp.61–77.

Walker, S.H., Carlisle, B.C. & Muddiman, D.C., 2012. Systematic comparison of reverse phase and hydrophilic interaction liquid chromatography platforms for the analysis of N-linked glycans. *Analytical Chemistry*, 84(19), pp.8198–8206.

Wang, W. et al., 2005. Proteome analysis of a recombinant *Bacillus megaterium* strain during heterologous production of a glucosyltransferase. *Proteome science*, 3, p.4.

Weber, B.H. et al., 1989. Evolution in thermodynamic perspective: An ecological approach. *Biology and Philosophy*, 4(4), pp.373–405. Available at: http://link.springer.com/10.1007/BF00162587.

Wendisch, V.F. et al., 2006. Emerging *Corynebacterium glutamicum* systems biology. *Journal of Biotechnology*, 124(1), pp.74–92.

van der Werf, M.J. et al., 2008. Comprehensive analysis of the metabolome of *Pseudomonas putida* S12 grown on different carbon sources. *Molecular BioSystems*, 4(4), pp.315–327. Available at: http://xlink.rsc.org/?DOI=b717340g.

van der Werf, M.J. et al., 2007. Standard reporting requirements for biological samples in metabolomics experiments: microbial and in vitro biology experiments. *Metabolomics*, 3(3), pp.189–194. Available at: http://link.springer.com/10.1007/s11306-007-0080-4.

Wiechert, W. & Noack, S., 2011. Mechanistic pathway modeling for industrial biotechnology: Challenging but worthwhile. *Current Opinion in Biotechnology*, 22(5), pp.604–610. Available at: http://dx.doi.org/10.1016/j.copbio.2011.01.001.

Williams, D.C. et al., 1982. Cytoplasmic inclusion bodies in *Escherichia coli* producing biosynthetic human insulin proteins. *Science (New York, N.Y.)*, 215(4533), pp.687–689.

Williams, J.C. & Weiss, E., 1978. Energy metabolism of *Rickettsia typhi*: Pools of adenine nucleotides and energy charge in the presence and absence of glutamate. *Journal of Bacteriology*, 134(3), pp.884–892.

Wilson, I.D. et al., 2005. High resolution "ultra performance" liquid chromatography coupled to oa-TOF mass spectrometry as a tool for differential metabolic pathway profiling in functional genomic studies. *Journal of Proteome Research*, 4(2), pp.591–598.

Winder, C.L. et al., 2008. Global metabolic profiling of *Escherichia coli* cultures: An evaluation of methods for quenching and extraction of intracellular metabolites. *Analytical Chemistry*, 80(8), pp.2939–2948.

Wittmann, C. et al., 2005. Dynamics of intracellular metabolites of glycolysis and TCA cycle during cell-cycle-related oscillation in *Saccharomyces cerevisiae*. *Biotechnology and Bioengineering*, 89(7), pp.839–847.

Wittmann, C., 2007. Fluxome analysis using GC-MS. *Microbial cell factories*, 6(6). Available at: http://www.microbialcellfactories.com/content/6/1/6.

Wittmann, C. et al., 2004. Impact of the cold shock phenomenon on quantification of intracellular metabolites in bacteria. *Analytical Biochemistry*, 327(1), pp.135–139.

Wittmann, C. et al., 2007. Response of fluxome and metabolome to temperature-induced recombinant protein synthesis in *Escherichia coli*. *Journal of Biotechnology*, 132(4), pp.375–384. Available at: http://linkinghub.elsevier.com/retrieve/pii/S0168165607009601.

Wittmann, C. & Heinzle, E., 2002. Genealogy profiling through strain improvement by using metabolic network analysis: metabolic flux genealogy of several generations of lysine-producing corynebacteria. *Applied and Environmental Microbiology*, 68(12), pp.5843–5859. Available at: http://www.ncbi.nlm.nih.gov/pubmed/12450803.

Wittmann, C. & Heinzle, E., 1999. Mass spectrometry for metabolic flux analysis. *Biotechnology and bioengineering*, 62(6), pp.739–750. Available at: http://www.ncbi.nlm.nih.gov/pubmed/10099575.

Wolff, J.A. et al., 1991. Isolation and characterization of catabolite repression control mutants of *Pseudomonas aeruginosa* PAO. *Journal of Bacteriology*, 173(15), pp.4700–4706.

Wu, L. et al., 2005. Quantitative analysis of the microbial metabolome by isotope dilution mass spectrometry using uniformly ^{13}C-labeled cell extracts as internal standards. *Analytical Biochemistry*, 336(2), pp.164–171.

Wu, L. et al., 2006. Short-term metabolome dynamics and carbon, electron, and ATP balances in chemostat-grown *Saccharomyces cerevisiae* CEN.PK 113-7D following a glucose pulse. *Applied and Environmental Microbiology*, 72(5), pp.3566–3577.

Xing, D. et al., 2008. Electricity generation by *Rhodopseudomonas palustris* DX-1. *Environmental Science and Technology*, 42(11), pp.4146–4151.

Yang, S., Sadilek, M. & Lidstrom, M.E., 2010. Streamlined pentafluorophenylpropyl column liquid chromatography-tandem quadrupole mass spectrometry and global ^{13}C-labeled internal standards improve performance for quantitative metabolomics in bacteria. *Journal of Chromatography A*, 1217(47), pp.7401–7410.

Yates, J.R., 2001. Mass spectrometry in biology. In *eLS*. Chichester, UK: John Wiley {&} Sons, Ltd, pp. 1–5. Available at: http://doi.wiley.com/10.1038/npg.els.0000999.

Yee, L. & Blanch, H.W., 1992. Recombinant protein expression in high cell density fed-batch cultures of *Escherichia coli*. *Bio/technology (Nature Publishing Company)*, 10(12), pp.1550–1556. Available at: http://www.ncbi.nlm.nih.gov/pubmed/1369204.

Yimga, M.T. et al., 2006. Role of gluconeogenesis and the tricarboxylic acid cycle in the virulence of *Salmonella enterica* serovar Typhimurium in BALB/c mice. *Infection and immunity*, 74(2), pp.1130–1140.

Yuan, J. et al., 2009. Metabolomics-driven quantitative analysis of ammonia assimilation in *E. coli*. *Molecular systems biology*, 5(302), pp.1–16. Available at: http://dx.doi.org/10.1038/msb.2009.60.

Al Zaid Siddiquee, K., Arauzo-Bravo, M.J. & Shimizu, K., 2004. Metabolic flux analysis of *pykF* gene knockout *Escherichia coli* based on ^{13}C-labeling experiments together with measurements of enzyme activities and intracellular metabolite concentrations. *Applied Microbiology and Biotechnology*, 63(4), pp.407–417.

Zakhartsev, M. et al., 2015. Fast sampling for quantitative microbial metabolomics: new aspects on cold methanol quenching: metabolite co-precipitation. *Metabolomics*, 11(2), pp.286–301.

Zamboni, N. et al., 2009. (13)C-based metabolic flux analysis. *Nature protocols*, 4(6), pp.878–892.

Zamboni, N., Kümmel, A. & Heinemann, M., 2008. anNET: a tool for network-embedded thermodynamic analysis of quantitative metabolome data. *BMC bioinformatics*, 9, p.199.

Zhou, X. et al., 2013. Observation of CO2 and solvent adduct ions during negative mode electrospray ionization Fourier transform ion cyclotron resonance mass spectrometric analysis of monohydric alcohols. *Rapid communications in mass spectrometry : RCM*, 27, pp.2581–2587. Available at: http://www.ncbi.nlm.nih.gov/pubmed/24591018.

Zurek, L. et al., 2001. Vector competence of Musca domestica (Diptera: Muscidae) for Yersinia pseudotuberculosis. *Journal of Medical Entomology*, 38(2), pp.333–335.

9 Appendix

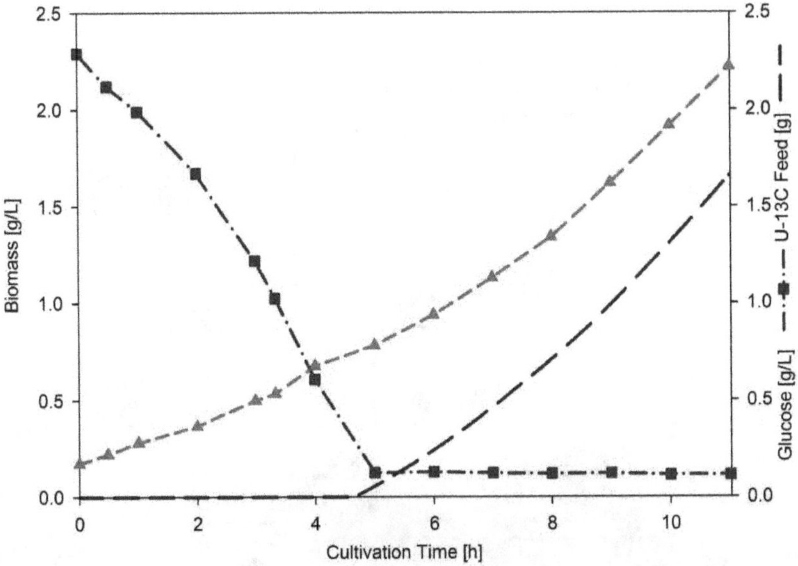

Figure 9.1: Development of glucose concentration and biomass during fed-batch cultivation of E. coli K12 DSM 2670 in a 1L-bioreactor filled with 0.4 L of minimal medium using [$^{12}C_6$] glucose as the carbon source. The feed solution contained [$^{13}C_6$] glucose. The applied feed profile supported a specific growth rate of μ=0.18 h-1 and was started at t = 4.3 h. Additionally, the total amount of added [$^{13}C_6$] glucose is depicted.

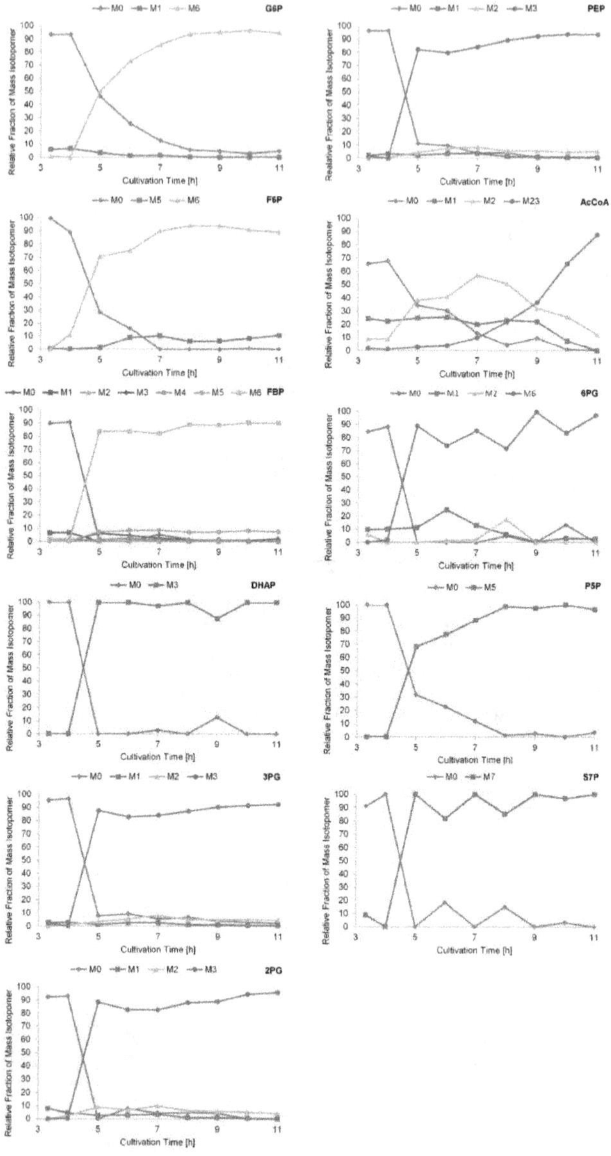

Figure 9.2: Differences in the ^{13}C labeling of metabolites from the central carbon metabolism of *E. coli* K12 DSM 2670 during the fed-batch experiment. The cells were grown in a 1L-bioreactor filled with 0.4 L minimal medium containing [$^{12}C_6$] glucose as carbon source. The feed profile supported a growth rate of 0.18 h^{-1}. The feed contained [$^{13}C_6$] glucose as the carbon source and was started as soon as the initial glucose concentration was depleted at t=4.3 h. Metabolites of glycolysis and PP pathway exhibited an immediate increase of the labeled fraction of metabolites (t=4.3 h). 100% of labeling was reached at t=8 h.

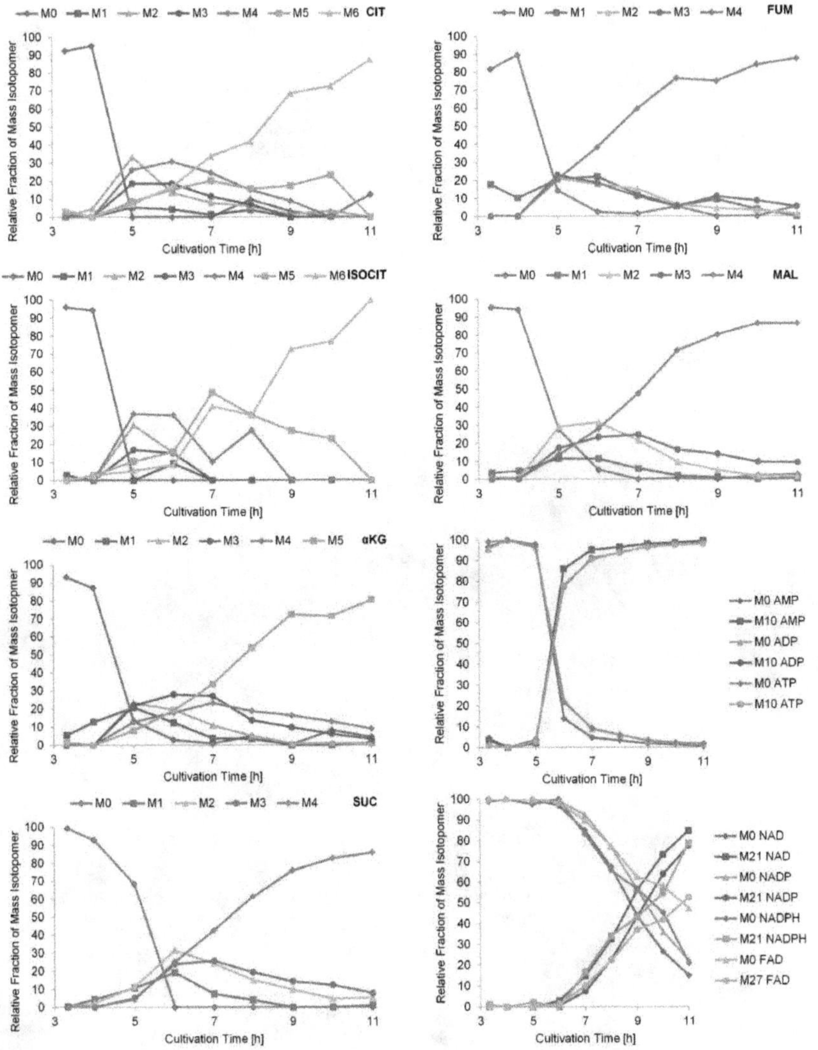

Figure 9.3: Differences in the ^{13}C labeling of metabolites from the central carbon metabolism of *E. coli* K12 DSM 2670 during the fed-batch experiment. The cells were grown in a 1L-bioreactor filled with 0.4 L minimal medium containing [$^{12}C_6$] glucose as carbon source. The feed profile supported a growth rate of 0.18 h^{-1}. The feed contained [$^{13}C_6$] glucose as the carbon source and was started as soon as the initial glucose concentration was depleted at t=4.3 h. Labeling of energy equivalents started 1 hour delayed (t=5 h), but was almost completely labeled within two hours (t=7 h). Labeling of redox equivalents began 2 hours after feed start (t=6 h) and was not completed at the end of the experiment. Labeling of metabolites of the TCA cycle could be seen immediately (t=4.3 h) but was not completed at the end of the experiment. Furthermore, various labeling states of the TCA cycle metabolites could be detected.

Figure S[illegible] [illegible] labelling of metabolites [illegible] the central carbon metabolism [illegible] [illegible] experiment. [illegible] grown [illegible] [illegible] profile [illegible] growth rate of [illegible] the [illegible] glucose as the carbon source [illegible] the initial glucose [illegible] depleted at [illegible] h. Labelling of energy [illegible] stem [illegible] was [illegible] labelled within two hours [illegible] began 2 hours after [illegible] and was [illegible] completed at the end of the experiment. Labelling of metabolites of the TCA [illegible] but was not completed at the end of the experiment. Furthermore, [illegible] metabolites could [illegible]

www.ingramcontent.com/pod-product-compliance
Lightning Source LLC
Chambersburg PA
CBHW060450240326
41598CB00088B/4333
* 9 7 8 3 7 3 6 9 9 9 7 1 8 *